The American Catholic

THE
AMERICAN
CATHOLIC

A Social Portrait

ANDREW M. GREELEY

Basic Books, Inc., Publishers

NEW YORK

To Senator Pat

Library of Congress Cataloging in Publication Data

Greeley, Andrew M   1928-
   The American Catholic.

   Includes bibliographical references and index.
   1.   Catholics in the United States.   I.   Title.
BX1406.2.G67   301.45'28'2   76-7683
ISBN: 0–465–00129–7 (cloth)
ISBN: 0–465–09733–2 (paper)

CONTENTS

ACKNOWLEDGMENTS

THIS BOOK represents the summary of fifteen years of research on the latter stages of the acculturation of the Catholic ethnic immigrant groups into American society. Various segments of the research described herein have been funded by the Carnegie Corporation of New York, the Ford Foundation, the Henry Luce Foundation, the John and Mary Markle Foundation, the National Conference of Catholic Bishops, and the National Institutes of Education.

I am indebted to many colleagues at NORC who have collaborated with me during the past decade and a half, and especially the four directors of NORC during that period, Peter Rossi, Norman Bradburn, James Davis, and Kenneth Prewitt. Other colleagues who have made notable contributions to my research are William McCready, Norman Nie, Sidney Verba, John Petrocik, Terry Clark, Christopher Jencks, James Coleman, Kevin Ryan, David Greenstone, Kathleen McCourt, and Harold Abramson. And beyond the limits of NORC, I am grateful for the helpful comments of David Riesman, Otis Dudley Duncan, Nathan Glazer, Daniel Patrick Moynihan, Seymour Lipset, and Arthur Mann.

Finally, I am also deeply indebted to Virginia Reich and Julie Antelman, who typed the manuscript and checked the references.

The American Catholic

Introduction

LET US BEGIN with a series of paragraphs describing the American Catholic:

1. Catholics tend to be blue-collar workers and belong to the lower-middle class. Educationally and financially they do not compare with their Protestant counterparts. A Catholic background makes it less likely that a young person will choose an academic career or, should he choose one, do well in it. Those Catholics who do become successful academics will leave the church. Irish Catholics, who have been in the United States longer than other ethnic groups, have achieved a certain amount of modest respectability, but they have not made the most of their opportunities in the new world—perhaps because of their religion, or their family structure, or their drinking habits.

2. While they have traditionally voted Democratic, Catholic ethnics are conservative. They are more likely to be racist, less likely to support civil liberties, more likely to take a punitive attitude toward the counterculture. They were stronger supporters of the Vietnam war than other Americans and voted heavily for Wallace in 1968. As many of them moved into the suburbs and became more affluent, they began to drift away from the Democratic party in both affiliation and voting behavior.

3. Most Catholic priests are not happy in their vocations. Those who have left the priesthood are the best and the most talented. The commitment to celibacy makes it impossible for a man to develop capacities of openness, intimacy, and sympathy with human frailty. Most Catholic clergy would marry given the chance.

4. While Catholics are more likely than Protestants to be against abortion, the opening up of the church in the Second Vatican Council has caused a notable decline in Catholic religious practice. The encyclical letter *Humanae Vitae* on birth control, however, caused serious moral anguish for large numbers of American Catholics.

5. Catholic support for their schools is declining, mostly because Catholics now realize that in the suburbs where they now live the public schools are better. There is little willingness in the Catholic population to make the financial sacrifices required to keep parochial schools in

operation. In any event, there is no evidence that parochial schools make their graduates any more religious than they would have been had they gone to public schools. Parochial school graduates are more likely to have hostile attitudes toward blacks and Jews, and less likely to be well equipped for success in American society.

All five of the above paragraphs seem unexceptionable. They are a fair portrait of what everyone knows to be true about certain aspects of American Catholicism. Why would one waste time, effort, and money to collect social science data to support such obviously true statements?

In fact, every single one of the above propositions is demonstrably false. There exists empirical evidence to demonstrate that each one is not true, and that in many cases the opposite is true. For example: The Irish are the most successful gentile group in the United States both financially and educationally. Support for Catholic schools has not declined in the last ten years, and there is considerable willingness to provide them with financial support—and this is as true for Catholics in their twenties as it is for those in the older age cohorts. Catholics did not support Wallace in any appreciable numbers. They have not drifted to the right politically, and they have not left the Democratic party. Catholics are no longer disproportionately blue-collar workers. On the contrary, they are more likely than Protestants to be members of the middle class (taking into account race, region of the country, and urbanism). Most Catholic priests are happy in their work, and only a minority would marry if they were free to do so. Psychological testing does not show them to be deficient in their capacity for intimacy. I could go on.

This book is intended to be a social portrait of American Catholics to be read mostly by non-Catholics. I would very much like to write an objective, dispassionate, and scholarly work, but I am afraid that will not be possible. Objective statements about American Catholicism based on the best survey data available become polemical by their very nature, because the truth about American Catholics is something that a considerable number of their fellow citizens, particularly those who constitute the country's intellectual and cultural elite, have not yet been able to understand or believe. But for those who will read what follows with an open mind, let me make four assertions which will set the context for and will in part be supported by the evidence presented in this volume:

1. Catholics in the United States are victims of what the anthropologists call a "cultural division of labor." Others would call this sophisticated discrimination.
2. Nevertheless, this discrimination has not prevented Catholics from achieving a substantial measure of financial success in the United States. It has been a less punitive discrimination than for blacks, let

us say. It may be roughly comparable to that aimed at white southerners in certain northern intellectual circles.

3. Hence, most Catholics are quite unaware of this "cultural division of labor" and are not disturbed by it. Paradoxically, however, precisely because of their success, dissatisfaction with a cultural division of labor is likely to increase.

4. Also paradoxically, it will increase precisely at a time when, because of incredibly inept administration, the organizational loyalty of a considerable number of Catholics is declining sharply.

The first assertion will be documented to some extent in chapters 1 and 3. Assertions 2 and 4 will be discussed in great detail in chapters 2 and 6; assertion 3 will be discussed in chapter 11 on neighborhood and in the concluding chapter. But here I wish to discuss in broad terms and in an introductory sense what I mean by the term "cultural division of labor" and its implications of discrimination toward American Catholics.

Michael Hechter, in his book *Internal Colonialism*, describes the "cultural division of labor" as a system of stratification in which one group, or a number of groups, is systematically excluded from certain key positions in society.[1] Hechter characterizes these groups as persisting in objective cultural distinctiveness and maintaining an assigned (by the core culture) and self-conscious "ethnic identity." He quotes Frederik Barth: ". . . ethnic identity implies a series of constraints on the kinds of roles an individual is allowed to play, and the partners he may choose for different kinds of transaction. . . ."[2] Hechter is writing about a situation in which this subordinate ethnic group is a victim of massive and pervasive discrimination. Such is *not* the condition of American Catholics. (Hence our propositions 2 and 3.) But it is true that there are certain kinds of positions in the United States in which Catholics are rarely represented—including its presidency until a decade and a half ago. There has never been a Catholic secretary of state or, in this century, a Catholic Supreme Court chief justice. Only two Catholics have ever been presidents of major American universities (Columbia and City University of New York). Catholics are virtually invisible on the boards and staffs of major foundations and on the tenured faculties of the arts and sciences divisions of our country's great private colleges and universities. Despite some very conspicuous exceptions (the last three executives of General Motors and the present executive of Ford Motor Company), Catholics are underrepresented on the boards and in the upper levels of corporate administration.[3] In particular they do not hold many important positions in the influential journals and networks of the national media, and are not actively recruited by local television stations in areas where there are large Catholic populations. Thus in Chicago, with its huge Slavic and Italian populations, the four television channels are very careful to recruit blacks, women, and Latinos for their staffs;

but one sees no Polish or Italian names, and when one asks where they are, one is told (as I was), "We looked for a Pole; we just couldn't find one who looked well on the tube."

In particular, those Catholic ethnic groups which migrated to the United States in the late nineteenth and early twentieth centuries are simply not to be found in the media, the arts, the corporate board rooms, the foundation staffs, and the great university academic departments— with some notable exceptions (especially Italian movie directors).

There is little disposition to deny the facts of this cultural division of labor. When it is called to the attention of foundation executives or university presidents that there are very few practicing Catholics around, they shrug their shoulders and give much the same reply as did the Chicago TV recruiter. Catholics don't seem to be interested in the work, they're not ambitious enough, or well educated enough, or intellectual enough—maybe it's something in their religion or family life, or both.

But I certainly have no desire to turn this book into a running argument with the reader. I intend to present a social description of American Catholics based, in almost all instances, on a solid national sample of empirical data. When the data are less than satisfactory, I shall say so. I will not attempt to make any recommendations or to argue for any programs, but I must warn the reader in advance that the findings on which I report—the results of fifteen years of social research—may be very hard for him to accept. They will be to a substantial extent at odds with the preconceptions, assumptions, stereotypes, and simple explanations that have for so long characterized the image of Catholics in the United States.

I do not enjoy questioning other people's assumptions. If I do so in this book it is only because any objective statement on the present condition of American Catholics cannot help but challenge assumptions that are powerfully held and that call into question pictures deeply embedded in the unconscious of many Americans—Catholics as well as non-Catholics.

My goal in this volume is sociological, not theological. I do not intend to present a summary of American Catholic doctrine. Nor do I intend a systematic presentation of what American Catholics believe, although I shall have some things to say about how some beliefs are changing. I am primarily concerned with the Catholic people and not with the Catholic church as an organization. Obviously the distinction between the two is not perfect; both inside and outside Catholicism the distinction often is not even taken seriously. Catholics have only one formal organization, the church—with the result that religious leaders have also served as social, educational, and political leaders when the formal concerns of the church were at stake. Catholic laymen who became political leaders had an immense personal influence, of course, but only through their political office. Within the Catholic community there is no social role for the independent lay activist such as those provided for the Jewish community by organizations like B'nai B'rith or the American Jewish Committee. One of the results of this lack of leadership differentiation is that Catholics, in their explicitly Catholic concerns, have been far more dependent on the skills of their religious leaders than any other major group within the society. When these leaders are inept—as they have been for the last twenty years—there is no one else to whom Catholics can turn for more effective leadership. Interestingly, one of the results of the decline of organizational loyalty to the church as such is the beginning of an organizational pluralism within the Catholic collectivity. Not surprisingly, some of the leaders of these new organizations (such as the National Conference on Ethnic and Neighborhood Affairs) are themselves priests, dissatisfied with the ineptitude of their official leaders.

Most of recorded Catholic history and a good deal of Catholic sociology are *institutional* history and sociology—biographies of church leaders, stories of the development of particular dioceses, studies of parishes or religious communities. If one wants to know what Catholics are thinking, one turns to what their church leaders are saying, so close seems to be the identification of the collectivity with the institution. In fact, this identification was never as complete as it may have seemed, and is gradually becoming much less important. In this book I intend to turn the matter around and to talk about American Catholics as a collectivity, discussing the church only insofar as it impinges on the development of a collectivity (principally in chapters 7, 8, and 9, which deal with the Catholic schools, the priesthood, and the changes in the church in the last ten years).

Non-Catholic Americans may know something about the theology of Catholicism or about its canonical structures; if they do not, there are certainly books available that will tell them. But most non-Catholics know very little about the educational, occupational, and economic status of Catholics, or about their family structures, their moral values, or their political, social, and racial attitudes and behavior; nor are they well informed about the impact of Catholic schools on society or the attitudes of Catholics toward parochial schooling. It is to fill all these gaps that this volume has been prepared.

REFERENCES

   1. Michael Hechter, *Internal Colonialism: The Celtic Fringe in British National Development, 1536–1966* (Berkeley, Calif.: University of California Press, 1975).
   2. Frederik Barth, "Introduction" in *Ethnic Groups and Boundaries*, ed. F. Barth (Boston: Little, Brown, & Co., 1969), p. 11.
   3. Russell Barta, "Representation of Poles, Italians, Latins, and Blacks in the Executive Suites of Chicago's Largest Corporations," *Minority Report* (Chicago: Institute of Urban Life).

CHAPTER

1

Models for Viewing
American Catholicism

THIS PORTRAIT of American Catholics is essentially a study in acculturation—in the broadest sense of the word. It is an attempt to measure how Catholic immigrants and their descendants have adjusted to a society which already had its own culture, social structure, and politics at the time the immigrants began to arrive. American Catholics are an ethnic group, in the general sense of the term, as well as being a group of ethnic groups, in the specific use of the word "ethnic" as meaning descendants of European immigrants.

The Catholic collectivity is a group which is in some respects different from the host culture. There are boundaries between Catholics and others in the United States. These boundaries are not legal, for the most part, although there are boundary-setting consequences of laws or judicial decisions. The Supreme Court decisions on Catholic schools, for example, set legal boundaries around those schools. Court decisions on "integration" (the so-called "affirmative action") in many cases discriminate against Catholics, because in many large urban centers discrimination in favor of nonwhites inevitably becomes discrimination against Catholics. However, the boundaries among religious groups in American society are cultural and social for the most part; they are implicit and unofficial.

Few subjects in American social science are more likely to stir up emotions and moral passions than ethnicity and acculturation. The pictures one carries around in one's head which deal with such subjects are not only descriptive—they are frequently prescriptive. One feels that one knows not only what has happened but that what has happened should have happened. Any attempt to question the factuality of the process is morally offensive. The moral passions which led to the restrictive immigration legislation in 1920 were based on a broad national consensus that the goal of acculturation was the assimilation of immigrant groups so that they lost their identities, and that the masses of southern and eastern European Catholic immigrants who were pouring into the society would be very difficult, if not impossible, to assimilate. The national consensus after World War I was xenophobic. (Early immigration laws required only an oath of allegiance, and the English language was not required for citizenship prior to 1907.) The strange, dark-skinned, oddly named immigrants from eastern and southern Europe were perceived as a threat to society. Their assimilation was considered a major challenge which had to be met if the society was to survive and remain healthy, and the public high school in particular was touted as the principal agent of assimilation. It was to undo all the bad habits and strange, crazy customs the children of immigrants picked up in their family environments.

The restrictive immigration legislation has been repealed, but the model of society which underlay it is still quasi official: children and grandchildren of the last waves of immigrants ought to become indistinguishable from "everybody else." A somewhat less rigid model is the "melting pot." In the "melting down" that was supposed to take place in the public cauldron, some of the characteristics of the more recent immigrant groups would be lost, others would be preserved to become part of the common heritage. Similarly, one supposes, in any effective melting pot, some of the characteristics of the host culture would be burned out, too. But the "official" assumptions of the last half century have certainly been that if anyone was going to change at all it would be the immigrants.

At a higher level of ethnic generalization, Catholicism was always viewed with dark suspicion by a sufficiently large group in the society to make it a convenient scapegoat for many social ills—both real and imagined. In the early years of the century it was the unknown dark southern and eastern European Catholics who were suspected of bringing dangerous radical ideas to the United States, and during World War I it was the largely loyal and partly Catholic German group that was thought to be treasonable. Much of the xenophobia of Attorney General A. Mitchell Palmer's crusade in the early 1920s paralleled the resurgence of anti-Catholic nativism in the post-World War I Ku Klux Klan, and both were partly manifestations of the more serious xenophobia against groups which "just happened" to be Catholic that was apparent in the report of the Dillingham commission. Paradoxically, a mere thirty years later, during the McCarthy era, the intellectual and cultural elite of America fantasized exactly the opposite position for Catholics. Instead of being the appropriate objects of the "red scare," they were among its principal advocates. Similarly, in the late 1930s and the early 1940s, when the official liberal position was interventionist, the Catholic groups were denounced for being isolationist; in the 1960s, when the "liberal" position shifted to isolationism, the Catholic ethnics were assumed to be the "middle Americans, the hardhat hawks who pushed the war in Vietnam.

For none of these "pictures" of Catholics was there any very strong empirical evidence; indeed, both the radical right of the McCarthy era and the hawks of the Vietnam war were not disproportionately Catholic. On the contrary, as we shall see in a subsequent chapter, James Wright's research shows that from the beginning Catholics were more likely than other Americans to oppose the Vietnam involvement. Still, the power of assumptions, particularly when they take the form of symbolic pictures, is such that they are not really subject to proof or disproof by empirical evidence.

Catholics in general have been viewed with suspicion since the beginning of their arrival in any numbers, not only by the masses but also by the cultural elites. Catholics were tolerable, perhaps, but disturbingly different. In the ideal order of things Catholics would become more like everyone else; through the passage of time and the influence of education, the differences which separated Catholics from other Americans would be melted away. Their clannishness, their divisive school system, their rigid ecclesiastical "discipline" (to use a word Justice Powell resurrected from the 1920s), their unmarried clergy, their large families, their peculiar religious practices—all of these would gradually disappear. By the beginning of this century the children of Massachusetts no longer had to go to school on Christmas day because there was no longer suspicion of this celebration as a "popish" feast; but for the first six decades of this century it was not clear that a Catholic could be elected president. Catholics, it was argued, were sufficiently different from other Americans that their "foreign allegiance" made it impossible for them to serve in the highest office in the land. When a Catholic was finally elected president, it was despite the loss of several million votes because of his religion.

Perhaps one of the major reasons for the resurgence of anti-Catholicism in the late 1960s and early 1970s in the United States was frustration over the fact that the ecumenical movement, as endorsed by the Second Vatican Council, seemed to promise the end of Catholic

strangeness. But Catholics kept their schools and their "divisive" attitudes on abortion. They did not stop being "strange" as the price for
ecumenical dialogue. They have not yet been assimilated.

So the Catholic collectivity is an ethnic group as well as a congeries
of smaller ethnic groups. Catholics constitute a group within the larger
society, and are perceived both by others and by themselves to be different to some extent. Part of the official "melting pot" model of acculturation to American society is that the differences *ought* to go away.

But what is an ethnic group, and what is ethnicity? Definitions
abound. To a considerable extent the definition one chooses is a function
of the perspective from which one is making an analysis or the point one
wishes to make. For my present purpose an elaborate but sufficiently
general definition will be adequate:

. . . a collectivity within a larger society having real or putative common
ancestry, memories of a shared historical past, and a cultural focus on one or
more symbolic elements defined as the epitome of their peoplehood. Examples
of such symbolic elements are: kinship patterns, physical contiguity (as in
localism or sectionalism), religious affiliation, languages or dialect forms, tribal
affiliation, nationality, phenotypical features, or any combination of these. A
necessary accompaniment is some consciousness of kind among members of
the group.[1]

The "pictures" we carry around in our heads help us to organize
and interpret our reality. Any comment on human society, whether it
results from common sense or rigorous scholarly research, also utilizes
pictures, perspectives, or paradigms. Such tools are indispensable for
the beginnings of analysis. David Matza comments on the functions of
these "pictures":

Pictures are intimately related to the explanation of social systems. Systems of
action may usually be typified in ideal fashion. Indeed, this simplification is
almost mandatory if the analyst wishes to proceed to the task of explanation.
A system, whether it be capitalism or delinquency, has exemplars, basic figures
who perpetrate the system. The accurate characterizing of exemplars is a
crucial step in the development of explanatory theory. Given the present state
of knowledge, pictures are not true or false, but rather plausible or implausible.
They more or less remind us of the many discrete individuals who make up a
social category.
    Systems of action have exemplars, and a portrayal of them is a crucial
step in the elaboration of casual theory. Thus, for example, a plausible picture
of the capitalist was implicit in the various theories explaining the rise of
capitalism. This hardly means that a system may be reduced to the character
of its exemplars; rather, an exemplar is a personification or microcosm of the
system. A crucial step from a Marxian to a Weberian theory of the origins of
capitalism consisted of a basic shift in the portrait of the exemplary capitalist.
Somewhere in the dialectic between competing scholars the pirate capitalist of
Marx was transformed to the bookkeeper capitalist of Weber. The more
authentic ring of Weber's portrait is largely responsible for the more widespread acceptance of his rather than Marx's theory of the emergence of

capitalism. Whatever the other virtues of Marx's theory, it suffers from an initial implausibility. It seems conceived on a false note. How, we ask, can we believe in a theory that apparently falsifies the character of the exemplars? Whatever the failings of Weber's theory, it seems more plausible because it is more reminiscent of the early capitalists we have studied or read about.[2]

The assumptions contained in such pictures are both absolutely essential for social research and dangerous for its goals: ". . . they tend to remain beyond the reach of such intellectual correctives as argument, criticism, and scrutiny. . . . Left unattended, they return to haunt us by shaping or bending theories that purport to explain major social phenomena."[3]

Two pictures shape most analyses of the ethnicity phenomenon. The first is the domination or oppression picture, the one Michael Hechter addressed himself to in his book, *Internal Colonialism*.[4] The second picture is the assimilation or, in its more popular form, the "melting pot" model, which I mentioned earlier. The domination image is basically a conflict picture in which one group is perceived as controlling and usually oppressing another. The assimilation image is more accommodationist in its assumptions and focuses on the adjustment of two cultures to each other after encounter and interspersion. Its concerns are "culture contact," in which the culture of the host society may be threatened by the culture of a numerically inferior but politically dominant group. It can also be the reverse phenomenon, in which the culture of the immigrant group is threatened by that of a numerically and political dominant host society. Both pictures assume a strain toward homogenization—political and structural in the domination picture and social and cultural in the assimilation picture.

In American society both pictures have been used, with the domination perspective applying mostly to relationships between white and nonwhite and the assimilation perspective applying mostly to relationships among the various white groups. It is the inadequacy of the latter picture that concerns me in this chapter.

The literature on assimilation in America is immense. Some authors see the process as rapid, others see it as slow. Some think it desirable that ethnic differences be eliminated so that a "common culture" may emerge, others think that assimilation should be decelerated so that many different cultures may flourish under the American umbrella. Milton Gordon combines the two by distinguishing between "structural assimilation," in which ethnicity is no longer pertinent even to primary group formation, and "cultural assimilation" (or acculturation), in which cultural differences diminish but the propensity to choose primary group relationships from within one's own group persists. Gordon argues that the latter process is far advanced in American society, while the former proceeds much more slowly.[5]

What all the assimilationist literature, popular and serious, sophisticated or simplistic, assumes is that the strain toward homogenization in modern industrial society is so great as to be virtually irresistible. The influences of the common school, the mass media, common political and social norms, and ethnic and religious intermarriage work toward the elimination of diversity in a society. Basic beliefs, socialization styles, personality characteristics, political participation, social attitudes, expectations of intimate role opposites, all tend toward a similarity that is differentiated only by social class. Social class is generally assumed to be a "rational" basis for differentiation, as opposed to differentiation based on religion and national origin, which are "irrational." Race was formerly an irrational focus for differentiation but is now rational.

The picture of American society as stated in the abstract categories of social science or the concrete categories of popular journalism is one of many different cultures merging into one common "American" culture. Only minor differences (such as special foods) persist. It may be debated whether this merging produces a totally new culture that is a combination of its various inputs, or whether in fact it is rather a matter of the various immigrant cultures adapting themselves to the host culture, which in the case of the United States I term "Anglo conformity." Whatever theoretical position one may take, once one assumes, as most of the literature does, that the immigrant culture is the dependent variable and the host culture the independent variable, Anglo conformity has entered the model.

The assimilation picture is pervasive in American society. It is part of our popular folk wisdom as well as an important component of the repertory of pictures available to social science theorists. Politicians, television commentators, movie critics, social planners, and reform political candidates all take it for granted. The picture has been wedged into our individual and collective unconscious, and has achieved the status not merely of conventional wisdom but of common sense.

Two things happen almost inevitably when such a picture becomes common sense. It becomes, as Matza suggested, undiscussed and undiscussable; it begins to become normative. It is now no longer a description of the way things are, but a description of the way things ought to be. Data that do not fit the picture are ignored, discarded, or subjected to the sorts of paralyzing questions against which no data can stand. Instead of being viewed as new and potentially informative findings, such data are written off as irrelevant or even as potentially dangerous.

It is but a short step from being undiscussably *descriptive* for a picture to become *prescriptive*. The picture becomes not merely an ideal type, it becomes a norm. To untangle the strands of nativism, liberal optimism, vulgar Marxism, secular rationalism, and immigrant selfrejection that underpin the melting pot norm is a challenge to which practitioners of the sociology of knowledge might want to respond. Sociology is supposed to involve the questioning of assumptions (even criticizing them, if we are to believe the younger members of the profession). Few have asked whether there might be other ways of looking at the phenomenon of ethnic differentiation in American society besides the "official" assimilationist picture or the moderately revisionist version advanced by Milton Gordon.

The fundamental assumption of the assimilationist picture is that in a modern society the forces working for homogenization—at least within broad social class groupings—are so powerful as to be irresistible. This assumption has been so deeply embedded in the collective unconscious that, with the exception of research by Harold Wilensky,[6] it is almost never questioned. But one need only look up from one's computer output, one's mathematical model, or the latest issue of the *New York Review of Books* to realize that differentiation runs rampant in American society. Processes of homogenization and differentiation are going on simultaneously. We are, to put the matter in popular terms, becoming more like one another and more different at the same time. A repertory of pictures of social reality that does not have room for paradoxical models may be neat, clean, and simple; whether it is helpful for understanding human behavior is another matter.

The assimilationist perspective is indispensable for coping with the social reality of America. The Irish ethnic and the Polish ethnic who live next door to one another have far more in common than their grandparents did—common language, common citizenship, a common set of television channels—but in some ways they may be more dissimilar. Their grandparents were in all likelihood peasant farmers, but the two American ethnics may have totally different occupational perspectives. Certain differences rooted in historical heritages may persist between the two Americans with no signs of diminution. For example, more than a member of any other ethnic group, the Irishman is likely to be a political activist. It is at least a researchable question as to whether in some respects the two neighbors are becoming increasingly different from one another. The Irishman may be defecting to the Republican party, while the Pole remains loyal to the Democrats. The assimilation picture alone will simply not do justice to this complex reality.

Certain limitations of the assimilation-acculturation picture must be considered. It frequently turns out to be not particularly helpful in generating hypotheses or in ordering data. It is difficult to determine, for example, whether a set of findings shows a high or low rate of acculturation. It also offers no insight into why there are presently some self-conscious attempts to create ethnic groups. In the Northeast, for example, there is a deliberate attempt to create a Spanish-speaking ethnic group (an attempt that is not supported, incidentally, in the Southwest).

An American Indian group is struggling to emerge, and in Chicago there is even an effort, as yet rather ineffective, to create an Appalachian white ethnic group. Harold Cruse has also suggested that the black power movement is essentially an attempt to create a black ethnic group,[7] a suggestion that Paul Metzger has echoed from a very different perspective.[8] The political and social leaderships concerned with the creation of ethnic groups must have insights into how power is exercised in the United States that are quite foreign to the acculturation picture.*

Similarly, the acculturation perspective does not take into account the fact, noted by many historians, that ethnicity was perceived by the immigrants as a way of becoming American. The hyphen in the hyphenate American was a symbol of equality, not inequality. In an urban environment where everyone, including the native American, was a something-else-American, one had to be ethnic to find one's place on the map. Thomas Brown notes that the principal argument of the nineteenth- and early twentieth-century Irish American nationalists who favored freedom for Ireland was that only when Ireland was a free and independent member of the family of nations would Irish Americans be accepted by the native Americans as worthy of full-fledged American citizenship. And Victor Greene has demonstrated that support among Polish Americans and Czech Americans for the nationalist movements in their native countries during World War I came only after the United States entered that war.[10] Such support for free Poland and the new Czech republic was, paradoxically, an exercise in American patriotism more than an expression of Polish or Czech patriotism.

More historical research is obviously required, but there is sufficient reason to state, at least as a tentative hypothesis, that the creation of ethnic groups in the United States was a way for the immigrant population to look at its present and future in America rather than at its past in the Old World. In a complex society of an "unstable pluralism"[11] you had to be "something" if you were going to be "anybody." Such a view of social reality is obviously foreign to the acculturation picture.

The acculturationist assumes that unstable pluralism is a socially dangerous situation. He expects social harmony to emerge out of the creation and reinforcement of a "common culture." In such a perspective the question becomes, how does a common culture emerge and how does it survive the assaults of periodic regressions to a rational differentiation? One who operates in the acculturation perspective has no way of

---

* Nathan Glazer has observed that the ethnic group came into existence in the United States.[9] It might be a mistake to conclude that the self-conscious formation of an ethnic group is a new development in American society. One wonders what reason there is to think that the leaders of previous efforts at ethnic group construction were acting unself-consciously.

addressing the question of how American society manages to keep from tearing itself apart despite its condition of unstable pluralism. During the last quarter of a century, when ethnic, racial, and religious violence has erupted all over the world, the United States has been relatively free from serious disruption. The urban riots of the late 1960s were minor in comparison with those of Indonesia, Bangladesh, the Sudan, Burundi, and Ulster. Despite the unstable pluralism which worried Americans in 1700, as Michael Kammen tells us, there has been only one civil war in this country, and that was mostly between British American groups. The pertinent question ought to be concerned with not how one protects and propagates the "common culture," but rather what there is in the national culture that has legitimated considerable diversity while creating at the same time implicit protocols by which most violent social conflict has been avoided. The assimilationist perspective marvels at how homogeneous American society is becoming and hence sees no real need for striving to understand the nation's capacity for observing and coping with our complex racial, religious, national, geographic, and social differentiation, a differentiation that came into being in a relatively short period of time in relation to the histories of human societies.

Another problem with the acculturation picture is that it does not account for the self-conscious manipulation of ethnic symbols in American society, a manipulation which ought to be increasingly difficult and infrequent but which in the social reality around us does not seem to be difficult at all. Polish and Italian self-consciousness, for example, can easily be written off as a response to black militancy. Yet it could also be argued that ethnic consciousness is merely a result of the fact that by accepting black self-consciousness the larger society legitimated public manipulation of ethnic symbols, which in prior years had been manipulated privately. A particularly interesting example is the appearance of tricolor bumper stickers on the cars of many Italian Americans in the eastern United States. It is safe to assume that most of these self-conscious Italians came from southern Italy and Sicily, where until fairly recently the tricolor represented the "foreign" domination of the Piemontese. The Sicilians came to the United States and discovered that they were Italian Americans. Now they have discovered they are Italian, a process exactly the reverse of that suggested by using only the acculturation picture. It is, of course, a research question as to how widespread the response is to such symbol manipulation. One would presume that sociologists would abstain from dismissing it as an irrelevant and unimportant phenomenon until they have studied it in detail.

As an alternative and complementary perspective, I would suggest that ethnic groups come into being in the United States and have a "natural history." The study of their genesis and history, free of the dogmatic assumption that their destiny is obliteration, can be useful in approaching both the history and the sociology of ethnic differentiation in the United States.

One need not subscribe to Levi-Strauss's view of the binary differentiating propensities of the structure of the human mind (indeed, I do not subscribe to his view) to be aware that humans "code" reality by differentiation. As Gerald Suttles has remarked, one creates neighborhoods so that one may have a chart of the city and know where one is going to encounter role-opposites whom one can reasonably trust.[12] But the mental chart that divides the city into many different neighborhoods is only one of a considerable number of such charts that we carry around in our minds (not unlike Matza's pictures). There are microstructure charts that divide the family into parents and children, and children into boys and girls or "big kids" and "little kids," for example. And there are macrostructure charts that divide the world or the human race or the population of the nation into various categories. Such categories enable us to engage in preliminary "coding operations" that help us to move in a tentative fashion through the maze of potential relationships that constitute human society. Even academics, proudly aloof from the prejudices and biases of ordinary men, still code their own departments by specialization and/or interaction networks, if not by cliques or factions. The pertinent question is not whether such charts exist but rather which chart will be imposed on what social phenomenon.* It is much like using overlays that drop down over the blank figure of the human body to show first the arterial system, then the venous, nervous, and muscle systems, until finally the picture is complete though incredibly complex.

For purposes of preliminary coding, such differentiation is not necessarily binary, although there may be a basic binary division of potential role relationships—those in which you can be reasonably trusting and those in which you must be cautious about extending trust, for example. Nor does differentiation need to be conflict producing. While it is certainly true that most conflict flows from some form of differentiation, it is not true that differentiation necessarily leads to conflict. Indeed, Rosemary Harris has pointed out that even in Ulster most Catholics and most Protestants are not in active conflict with each other, but rather live in conditions ranging from suspicious coexistence to reserved friendship.[13]

It is a truism that there are no native Americans (except the Indians, of course), but it is frequently overlooked just how recent this immigration was for many American families. Indeed, in the middle 1960s,

---

* One of my colleagues observed that when she was growing up in Florida she thought of herself as an American; when she went to Washington, D.C., she discovered she was Cuban; and when she came to Chicago she was told that she was "Spanish-speaking."

half the American Catholic adult population was either the first or second generation in this country. If one believes in the power of the mass media and the common school to wipe the cultural and psychological slates clean in the space of a few decades, the persistence of ethnicity will come as a surprise. But if one believes in a certain inertial strength of cultural traits and family memories, it is not at all surprising that collectivities which took their origin in a very recent immigration experience persist.

Historian Michael Kammen observes that American society was structured in its formative years by men who had a remarkably high degree of self-consciousness about the problem of cultural diversity, or, as Kammen calls it, "unstable pluralism":

And so it was that American colonial history, which had begun with a quest for purity and homogeneity, ended with a sophisticated rationale for pluralism and heterogeneity. What had happened was not really so paradoxical as it may seem, for the so-called melting pot had been a boiling cauldron all along, from Jamestown to James Madison. There is a very real sense in which the American nation emerged, not in response to new-found national unity, but rather in response to provincial disunity, in response to a historical problem of long duration: how best to control unstable pluralism, how best to balance the areas of compulsion and freedom in American life.[14]

However much they may have lamented it, those who created the structure of federalism had no choice but to build a political apparatus that could deal with "unstable pluralism." By the very fact that the structure was so tolerant of differentiation (although many American citizens were less so), there was considerable opportunity for later immigrants to create their own interest groups that would correspond to those they saw already existing in the society (not all of them social class groups). Madisonian federalism was designed, as Gordon Wood has suggested, to absorb the dynamisms of unstable pluralism by a process of government through compromise and coalition formation.[15] The immigrants soon discovered, perhaps unconsciously and unexplicitly, that to be a part of the coalition-formation game they had to have a collectivity of their own. Since the rest of society categorized people on the basis of nation of origin, it seemed sensible to go along with the process and use it for their own political, economic, and social advantage.

Ethnicity reveals itself as a relatively safe form of differentiation. While there may be conflict and competition among the various ethnic groups, the conflict is rarely violent. Society has implicitly legitimated ethnic differentiation (if not required it) and has provided protocols and processes whereby the potential conflict that could arise from such differentiation is minimized. The immigrants never saw their claim to be hyphenated Americans as involving any danger of tearing apart the new society, which on the whole was relatively benign to them. They may have been accused of being un-American on occasion and, more often, suspected of not being American enough, but they never realistically perceived themselves as a threat to the relative peace and harmony of the society. Despite the fears of the advocates of a common culture and assimilation, ethnic differentiation was never a serious threat to the social order.

Furthermore, it provided the immigrants and their children and grandchildren with considerable political, social, economic, and psychological advantages. The ethnic group became one of the avenues to political power for immigrants. It provided a special market in which the emerging business and professional class within the immigrant community could build its own economic base. It offered a social mobility pyramid that the more ambitious immigrants could ascend; if the social pyramid of the host culture was inaccessible, they could at least move to the social apex within their own collectivity. Psychologically it also provided continuity between the Old World and the New and made possible the preservation of a minimum of family values that were thought to be essential. To say that any of these functions has diminished importance for the children and grandchildren of the immigrants is to advance a hypothesis that has not been supported by research evidence.

The ethnic collectivity served as a context in which certain skills, traits, and characteristics brought from the Old World and proven advantageous in the New could be preserved and strengthened. Did hard work and intellectual ambition prove extremely helpful in American society? Then such work and ambition could be reinforced by telling children that it was an especially Jewish trait, and that to be good Jews they must develop it. Did a certain kind of informal political skill open up avenues to power and prestige? Then such political skills could be legitimated and reinforced on the grounds that they were particularly Irish. The conviction that a particular trait or style of behavior is characteristic of one's "own kind" legitimates that behavior style or trait, at least in one's own mind.

No differentiation is without cost, but the price of ethnic differentiation in American society is modest—if one is white. There are certain clubs, buildings, and companies from which Jews are excluded, for example. The psychological costs may be more severe. The emphasis on "respect" among some of the current militant ethnic groups indicates how pressing they may be for some people. Many ancestral memories had to be repressed (the Irish have been particularly successful at this repression) for one to become thoroughly American. But given the situations the immigrants left behind, the costs of ethnic differentiation seemed relatively minor compared to the benefits the new society was capable of bestowing on them, or, more precisely, benefits that could be wrested from the host society if one was prepared to join with one's own kind to create a more equal match.

Ethnic differentiation, then, turns out to be safe, to involve a number of important payoffs, and to be relatively inexpensive. Under such circumstances it would be surprising if ethnic groups did not emerge. Implicit in this perspective is the notion that ethnicity is not a residual social force that is gradually disappearing; it is a dynamic, flexible social mechanism that can be called into being rather quickly and transformed and transmuted to meet changing situations and circumstances. The coming into being and the transformation and transmutation of the ethnic collectivity constitute extremely useful foci for social research.

But why is ethnic identification not more important in American society than it seems to be? Why are the boundaries of the ethnic collectivity not more rigid? Surely ethnic identification is not nearly as decisive in the United States as it is, let us say, in the north of Ireland or in Belgium—to say nothing of Cyprus and Nigeria. On the contrary, the chart of differentiation provided by the ethnic overlay may be useful in American society, but the relatively recent immigrant quickly discovers that while there are ethnic boundaries in the United States, those boundaries are permeable—especially for someone who has made or is in the process of making a great deal of money. But whence comes this permeability?

It might be argued that the most striking aspect of American ethnic differentiation is that, by definition, ethnic boundaries are *supposed* to be permeable. In other countries ethnicity is considered a method of finding oneself in a systematic way as being "over against" the rest of society. Under such circumstances ethnic differentiation implies ethnic separation. But among white groups (and the theory becomes a dilemma on the subject of nonwhite groups) in the United States ethnicity has never been primarily a means of separation, much less isolation. As I remarked earlier, the combined form denotes equality; it is not a way of withdrawing from the rest of society so much as an institution for dealing oneself into it. The ethnic collectivity does indeed provide a rationale for self-definition, and implicit in all self-definitions is some sort of separation from those who do not share the same one. But ethnic self-definition in the United States is more concerned with defining oneself as part of the American society and not separate from it. Under such circumstances ethnic group boundaries are permeable because the political and social culture has decreed that they ought to be and because they have been so structured.

It is also required by both the national, political, social culture and by the implicit constitutional structure that membership in an ethnic collectivity be optional. One has the right, American society assumes, to be an ethnic if one wants to be, but one is under no obligation. In practice, of course, it is easier for some individuals to dispense with their ethnic identification than it is for others. Blacks, Chinese, Japanese, American Indians, to some extent Chicanos, and perhaps to a lesser extent Puerto Ricans and Cubans, would find it difficult to persuade other members of society that they are not part of the ethnic group to which they have been assigned. But in theory and to a considerable extent in practice the ethnic collectivity is a community of "limited liability" (to lift a phrase from a different but related discussion), one of the many such communities of limited liability that are available to an American. Whether and when a person chooses so to identify himself in his own thinking is completely up to him, and, in theory though scarcely yet in practice, such decisions ought to be accepted by others. To return to the image of the overlay, the ethnic chart is available to be used or not, frequently or rarely, whenever one wishes to code the possible relationships that are available.

I shall recapitulate here by introducing four schematic charts.* The reader must realize that sociologists are fond of such charts, and that we cheerfully concede that they are bare bones oversimplifications of reality. Nonetheless, as Otis Dudley Duncan has remarked, the most obvious use of such charts is that they force us to make explicit the implicit causal explanations that underlie our prose. These charts are designed to show the direction of influence on the culture systems of the immigrants through time and an increasing number of generations since immigration and experience in the common school. Each chart represents one of the perspectives on ethnicity discussed earlier.

Figure 1.1 presents the Anglo conformity perspective. Host and immigrant culture systems are separate. Through time, as the immigrants extend their generations in the United States and experience the influence of the common school (one might wish to include the mass media as "educational"), they become more and more like the hosts. Finally, at a certain moment either in the present or in time yet to come, host and immigrant are part of one common "American" culture that remains, however, the Anglo-American host culture.

Each box in the figure is a collection of myriad cultural variables. The figure suggests that in each of these variables the immigrants become more and more like the host as generation and education increase. The most obvious test of the validity of this perspective is whether, when generation and education are controlled, the differences between hosts and immigrants disappear. Since there is a strong correlation between education and generation, controlling for one is for all practical purposes

* For this discussion I rely heavily on my book *Ethnicity in the United States: A Preliminary Reconnaissance* (New York: John Wiley & Sons, 1974).

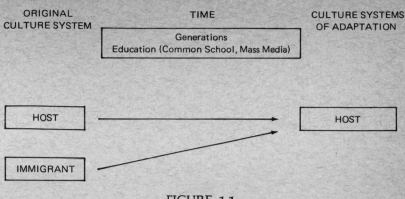

ORIGINAL   TIME   CULTURE SYSTEMS
CULTURE SYSTEM   OF ADAPTATION

Generations
Education (Common School, Mass Media)

HOST   HOST

IMMIGRANT

FIGURE 1.1
Anglo conformity perspective.

SOURCE, figures 1.1–1.5: Andrew M. Greeley, *Ethnicity in the United States: A Preliminary Reconnaissance*, pp. 304–309. (Reprinted by permission of the publisher, John Wiley & Sons—© 1974.)

the same as controlling for the other. Nonetheless whenever possible we take into consideration the impact of both education and generation. Unfortunately, many of the data sets available to us lack generational information. To the extent that one finds cultural variables in which the immigrants still differ from the hosts when education and generation are held constant, the Anglo conformity perspective loses some of its utility. But defenders of this perspective could argue that the amount of time allowed to effect assimilation is not sufficient. It is still taking place, they might contend, and will be accomplished in the future. One might respond, however, that if those groups who are well into their fourth generation continue to be different, the Anglo conformity perspective leaves much to be desired.

The "melting pot" usually means Anglo conformity. However, in its more romantic statements, such as in Israel Zangwill's famous play,[16] the concept has a slightly different meaning. In Figure 1.2 the movement is not completely of immigrant toward host but of host and immigrant toward each other, so that the common American culture that emerges is a combination of two cultures, though it is never clear how much and what the host culture absorbs from the immigrant. Such a model is difficult to test since we do not have measures on the host culture at the time a specific immigrant group arrives. Certain limited tests might be made, however. One could, with enough patience and resourcefulness, measure the diffusion of Italian food on menus in American restaurants in the past seventy years, or the diffusion of black slang expressions among those who speak white English, or perhaps the diffusion of the

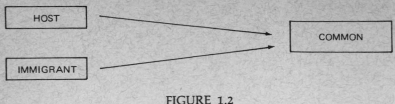

FIGURE 1.2

Melting pot perspective.

celebration of St. Patrick's Day. Clearly the creation of a common culture in which each group shares some culture traits of the others does occur in American society, but whether the melting pot is an important process in American society in this sense must be considered problematic.

A third perspective (Figure 1.3) is that of the classical cultural pluralism position as enunciated by Horace Kallen.[17] The immigrant does become like the host to some extent, hence the line jogs upward. He becomes an American citizen, commits himself to American political values, learns the English language, and enjoys the common mass media as well as the media of his own group. This is cultural pluralism as it exists in countries like Switzerland, Holland, Ulster, Ceylon, and perhaps some African nations.* However, few observers think that this kind of cultural pluralism exists in the United States, although certain black and Chicano groups advocate such a pluralism for their own communities. There are unquestionably small groups in the society, mostly rural (though the Hasidic Jews in Williamsburg are certainly urban), that have

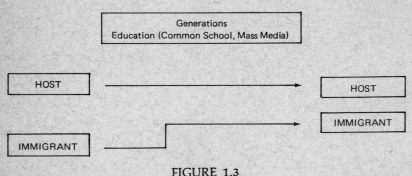

FIGURE 1.3

Cultural pluralism perspective.

* James Duran suggests that in cities in Kenya the ethnic diversity that emerges may be more like the American variety than the European. See "The Ecology of Ethnic Groups from a Kenyan Perspective," *Ethnicity* 1 (April 1974): 43–64.

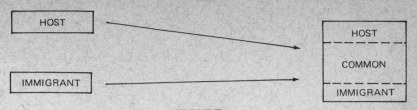

FIGURE 1.4

Acculturation but not assimilation perspective.

been able to sustain something very much like the classical picture of cultural pluralism.

When most contemporary defenders of ethnic diversity speak of cultural pluralism they mean something like the perspective presented in Figure 1.4. Milton Gordon has labeled this perspective "acculturation but not assimilation."* This position sees the immigrant absorbing large numbers of cultural traits from the host and the host picking up a few traits from the immigrant. What emerges is a common culture shared by both immigrant and host. But because "acculturation" occurs, it does not follow that "assimilation" does too. Some cultural traits still distinguish host from immigrant; in particular, the two groups maintain some distance from one another in the private spheres of their lives. Intermarriage does occur, but at a substantially lower rate than if the choice of marriage partner occurred independent of ethnic background. Similarly, close friends, recreation partners, and informal associates are far more likely to be chosen from one's own ethnic community than from the common pool of society. Put quite simply, when people are free to choose they choose from their own kind—even though they share large numbers of cultural traits with other groups in the society.

In a limited sense the fourth perspective is simplicity itself to test. One attempts to determine whether marriage is independent of religio-ethnic background. As Harold Abramson has demonstrated,[18] the ethnic factor is still pertinent in choice of spouse, although less than it used to be. One might also ask whether immigrant and host continue to be different in a certain number of cultural traits, and whether these differences persist within educational and generational groups. Those who doubt the utility of the acculturation-assimilation perspective might respond to such a test by saying that the difference which persist are declining, and

---

* Figure 1.4 is a simplification of Gordon's position in his book, *Assimilation in American Life* (New York: Oxford University Press, 1964). Gordon's position has become more refined and subtle through the years, and was presented in its most recent form in a paper at the American Academy Conference of Ethnicity, Boston, October 26–28, 1972. It appears in *Ethnicity; Theory and Experience,* ed. Nathan Glazer and Daniel P. Moynihan (Cambridge, Mass.: Harvard University Press, 1975).

in any event the variables being measured are not "important." Usually those who argue in such a fashion find the Anglo conformity perspective more to their taste.

The reader will have noted that I assume only two groups of cultural system, host and immigrant. Obviously there have been many immigrant groups, depending on time and place of arrival, and probably many host groups as well. There is no reason to assume that each immigrant group has moved along the lines of my figures at the same pace and in the same fashion. Their numerical size, education upon arrival, internal cohesion, resources provided them by their religion, skills, and sophistication of their leaders, among other factors, affect the process of merging with and adjusting to the host cultures. In an ideal world the four perspectives would be applied to each of the immigrant groups, and even then we would still have the problem of interaction of the immigrant groups with each other. For example, it is reasonable to suppose that German, Polish, and Italian Catholic groups experienced encounters not only with the dominant Anglo-American group but also, and perhaps more importantly, with the Irish Catholic group that ran (and to a considerable extent still runs) the Roman Catholic church.*

These figures, then, may be a somewhat pathetic attempt to schematize a reality of extraordinary variety and complexity, but they are a beginning. What we are suggesting is a picture of American ethnic differentiation that sees immigrants forming collectivities of limited liability based on presumed common origin, because these collectivities are tolerated and even encouraged by the larger society, provide substantial payoff with marginal risks, and incur only limited costs. In addition to studying the acculturation of immigrants, we should be also studying their ethnicization; that is, we should study the genesis and natural history of such collectivities. And we should be studying them in full consciousness that they are dynamic, flexible mechanisms that grow and change and whose disappearance ought not to be assumed on a priori grounds.

The fifth figure is a development of the four perspectives on ethnicity I presented earlier. It schematizes the ethnicization perspective, and is in some measure an extension of the acculturation-assimilation perspective shown in Figure 1.4. There are a number of important differences, however.

Figure 1.5 shows that the host and immigrants may have had something in common to begin with. Some of the Irish, for example, spoke English and understood something of the English political style of the eighteenth and nineteenth centuries. The other European groups were

---

* The Irish constitute 15 percent of the Catholic population, 30 percent of the clergy, and over half the hierarchy. See Andrew M. Greeley, *American Priests: A Report of the National Opinion Research Center* (Chicago, 1971).

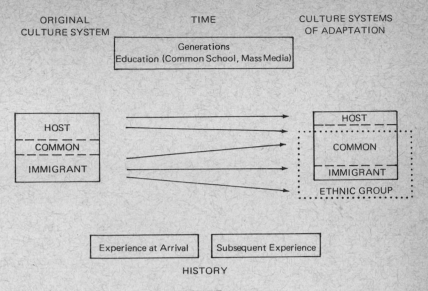

ORIGINAL
CULTURE SYSTEM

TIME

CULTURE SYSTEMS
OF ADAPTATION

Generations
Education (Common School, Mass Media)

HOST

COMMON

IMMIGRANT

HOST

COMMON

IMMIGRANT

ETHNIC GROUP

Experience at Arrival   Subsequent Experience

HISTORY

FIGURE 1.5

Ethnogenesis perspective.

part of the broad Western cultural inheritance. Under the influence of education, generation, and the experience in American society both at the time of immigration and subsequently, the common culture grows larger. Immigrants become more like the host, and the host may become somewhat more like the immigrants. Certain immigrant characteristics persist; in fact, under the impact of the experience of American life, some traits become more rather than less distinctive. Certain aspects of the immigrant heritage are emphasized and developed in response to the challenge of American society. What appears at the end (the right-hand portion of Figure 1.5) is that the ethnic group has a cultural system that is a combination of traits shared with other groups and traits that are distinctive to its own group. For the ethnics, then, the mix of traits and the emphasis within the cultural system are different from those of their immigrant predecessors. They share more with the common culture than they did to begin with, but in some respects may also be more different from the descendants of the hosts than their ancestors were from their hosts. In principle there is nothing to prevent testing of the various components of my perspective. In practice, however, an immense amount of social and historical research will be required. It is worth noting, incidentally, that while all the lines in Figure 1.5 are straight, in the reality this chart attempts to schematize the lines might well be jagged. For example, if one considers the variable of ethnic consciousness as part of the original immigrant system of traits, that consciousness may well have waxed and waned through the years, moving away from the common culture, then toward it, and away again in zigzag fashion.

In Figure 1.5 I have tried to combine the four previous perspectives on American ethnic diversity within a broader and more precise framework. I may have only complicated something already too complex, but the influence of the immigrant groups' experiential history in this country and in the country of origin toward the creation of distinct cultural systems is too important to ignore.

The two principal models for viewing the phenomenon of ethnicity and acculturation in the United States are the melting pot model of the assimilationists and the mosaic model of the cultural pluralists. Both have their uses, at least as descriptions; neither have been effective as prescriptions. In fact, I would suggest that the most useful model available to us is a mosaic with permeable boundaries. There are boundaries among the various nationality groups in the United States, there are boundaries among the religious denominations, there are boundaries between Catholics and non-Catholics; but these boundaries are not fixed or impermeable. One can choose which collectivity to affiliate with, and one can also choose the degree of affiliation. For example, one can be a "devout" Catholic, an "active" Catholic, a "practicing" Catholic, a "marginal" Catholic, or a "disaffiliated" Catholic, and one can move back and forth among the various positions along the affiliation continuum. One can be angry at one's Catholic past, ignore it, be militant about it, be fascinated by it, or endeavor to forget it completely (although Catholicism as a world view and Catholicism as a means of ethnic identification are both acquired very early in the childhood experience, and hence, both are very difficult to dispose of). American society does not force you to be or not to be Catholic; it leaves Catholicism open to you as a world view and also open as a means of self-definition. Catholicism is a collectivity, an ethnic group which is available for your use if you want to use it. It can be important to you in choosing friends, fellow workers, a professional career, and especially a marriage partner. Given the pluralistic nature of American society and the importance of religion and ethnicity in defining one's place within the society, most people tend, more or less, to identify with the religious or the religio-ethnic collectivity. There is no evidence that such a tendency is diminishing appreciably, but it is not nearly so rigid as the model of the mosaic society presents it. Precisely because it is such a flexible tool, the permeable boundary has a much better chance of surviving in a dynamic, mobile society.

The Catholic collectivity, then, as it exists in America, is to some considerable extent an American creation. It is a religious group which provides both world view and self-definition, and which passes on certain behavioral traits, values, and personality characteristics in a process which is normally quite unself-conscious. The Catholic collectivity is much more tightly defined here than it is in a country like Italy, where virtually everyone is Catholic in some fashion. But it is also much more loosely defined, in the sense that its boundaries are much more easily crossed than would be the case in countries like Holland or northern Ireland, where religio-ethnic boundaries are very sharp and rigid. In virtually all Western societies, denominational groupings have some sort of boundaries which make them at least quasi-ethnic groups; only the United States has the curious mixture of a high degree of religious or religio-ethnic self consciousness and relatively flexible and permeable boundaries. To be a Catholic means to be different. Catholicism offers you a potentially important form of self-definition, a social location in which you can define yourself by those very differences as being separate from the rest of the society, but the separateness need not and usually does not imply hostility or major conflict; it is rather a way of carving out a piece of social turf for yourself. It is ground on which to stand, but not a fortified area to defend. It is ground around which one can wander more or less as one pleases, so there is no reason for yielding this useful means of social location and self-definition as a price for participating in the larger society. On the contrary, it is viewed as a means of participating in the larger society.

The theoretical perspective for viewing American pluralism laid out in this chapter calls into question the notion that with the passage of time the differences between Catholics and other Americans will go away. Ethnic groups develop a momentum and a life history of their own, sometimes veering more toward assimilation, other times leaning toward more highly conscious differentiation. Simple unilinear models of the relationships among ethnic groups and between ethnic groups and the host culture are usually not very helpful. Indeed, as I shall argue at the conclusion of the book, the American Catholic collectivity is presently going through a period when much of its former organizational loyalty to the Catholic church as institution is waning (largely because of the incredibly inept leadership of this institution). But at the same time, its loyalty to certain "ethnic" manifestations of Catholicism, such as parochial schools, is not changing in the slightest, and a much more self-conscious and reflective approach to those aspects of the Catholic heritage, which hitherto had been a largely unself-conscious process, seems to be increasing. I shall argue in subsequent chapters that in socioeconomic terms the acculturation, particularly of eastern and southern European immigrants, has been a success beyond anyone's expectation (save that of the immigrants themselves). But this socioeconomic acculturation does not mean southern and eastern European groups are becoming like British Americans or like Jewish Americans. The Poles, once devout Catholics, may be less likely to go to church because of their socioeconomic acculturation (and the stupidity of church leadership); the Italians, once not particularly devout in observable religious practice, may now go to church more often (despite the stupidity of church leadership); and both Italians and Poles may become much more explicitly conscious of their ethnic and religio-ethnic background and more irate about how the larger society and the Irish-dominated institutional church have treated this background with contempt. So we can see how in one sense these groups may have become less Catholic, and in another sense more so. On some matters they are going beyond the permeable boundaries of the Catholic collectivity, but in other matters they are returning more self-consciously to the core of that collectivity.

The model for considering American Catholics and the various subgroups within Catholicism that has been presented in this chapter is not a simple one. The melting pot and the mosaic models are easy ways to conceptualize American society; they do not require complex, nuanced, and dynamic pictures. They are therefore easy to use, and just as easily deceptive. The mosaic with permeable boundaries approach is more difficult to use but subsumes more evidence. Minimally it prepares one to encounter surprising empirical evidence, which is perhaps the best criterion one can find for a useful social science model.

REFERENCES

1. Richard A. Schermerhorn, *Comparative Ethnic Relations: A Framework for Theory and Research* (New York: Random House, 1969), p. 123.
2. David Matza, *Delinquency and Drift* (New York: Wiley, 1970), pp. 1–2.
3. Ibid., p. 1.
4. Michael Hechter, *Internal Colonialism: The Celtic Fringe in British National Development, 1536–1966* (Berkeley, Calif.: University of California Press, 1975).
5. Milton Gordon, *Assimilation in American Life* (New York: Oxford University Press, 1964).
6. Harold L. Wilensky, "Mass Society and Mass Culture," *American Sociological Review* 29 (April 1964): 173–197.
7. Harold Cruse, *Crisis of the Negro Intellectual* (New York: William Morrow, 1971).
8. L. Paul Metzger, "American Sociology and Black Assimilation: Conflicting Perspectives," *American Journal of Sociology* 76 (January 1971): 644–647.
9. Nathan Glazer, "Ethnic Groups in America," in *Freedom and Control in Modern Society*, ed. Monroe Berger, Theodore Abel, and Charles H. Page (New York: Van Nostrand, 1954), pp. 158–172.
10. Victor R. Greene, "For God and Country: The Origins of Slavic Catholic Self-Consciousness in America," *Church History* 35 (December 1966): 446–460.
11. Michael Kammen, *People of Paradox* (New York: Knopf, 1972).
12. Gerald D. Suttles, *The Social Construction of Communities* (Chicago: University of Chicago Press, 1972).

13. Rosemary Harris, *Prejudice and Tolerance in Ulster* (Totowa, N.J.: Rowman and Littlefield, 1972).

14. Kammen, *People of Paradox*, pp. 73–74.

15. Gordon Wood, *Creation of the American Republic 1776–1787* (Chapel Hill: University of North Carolina Press, 1969).

16. Israel Zangwill, *The Melting Pot*, 2nd rev. ed. (New York: Macmillan, 1917).

17. Horace M. Kallen, *Cultural Pluralism and the American Idea* (Philadelphia: University of Pennsylvania Press, 1956).

18. Harold J. Abramson, *Ethnic Diversity in Catholic America* (New York: Wiley, 1973).

CHAPTER
2

Background

Most of the research done on American Catholicism is institutional, concerned with the church as an organization and not as a population collectivity.[1] From the institutional histories one can only get an indirect feel for the state of the American collectivity. To make matters worse, many of the books, particularly those written before 1960, maintain a certain pious respect toward their subjects for fear of shocking Catholic readers, and perhaps also for fear of providing ammunition to the enemies of the church. Thus John Tracy Ellis' life of Cardinal Gibbons gives no trace of the folk tradition that "Jimmy Gibbons was a cute one," meaning that the cardinal archbishop of Baltimore was an extremely devious ecclesiastical operator. Furthermore, Monsignor Ellis' account of the appointment of the first apostolic delegate, in the same book, would appear to take a benign view of this development. Ellis is also apparently responsible for the decision not to print the love letters written by John Lancaster Spalding, the bishop of Peoria, to Mary Gwendolyn Caldwell, founder of the Catholic University of América and his mistress of twenty years.

Nevertheless, a number of interesting themes emerge from a review of the institutional literature. First of all, until about 1875 there was extraordinary physical hardship and danger in being a missionary in the United States. The Mercy order's foundation in Chicago, for example, was wiped out in a cholera epidemic of 1854. Men were made bishops in their early or middle thirties simply because older men were not capable of sustaining the physical effort necessary to travel around the diocese (England, Gibbons, Keane, Spalding, Ireland—the greats of the

American hierarchy—were all bishops in their middle thirties). Second, until the development of the trans-Atlantic cable and the fast sailing vessel, Rome was a great distance from the United States, and it took the early bishops longer to get from Baltimore to Washington than it takes present-day bishops to go from Baltimore to Rome. Hence, while there was an eagerness to keep Rome happy, there was also a predisposition to go one's own way and shape ecclesiastical practices to meet the actual needs experienced rather than the theoretical requirements of canon law.

The great nineteenth-century leaders were all appointed before the development of rapid means of transportation and communication, and before the coming of the Apostolic Delegate at the end of that century. They were pioneers in the American sense of the word, but they were also intellectuals—reading, writing, and speaking far more than any bishops since then have done. Many of the suggestions they made for ecclesiastical structure and practice, based on their experiences as American bishops, would later be voted in by the Second Vatican Council (with most of the American delegates to that session unaware that they were supporting things their predecessors stood for).

Because of the vast reaches of America, the primitive conditions, and the distance from Rome, there was constant internal conflict within the church organization. Canon law provided no easy solutions for the problems which arose. The missionary clergy who were attracted to America were often unruly outcasts, and came from conditions where bishops had relatively little power over their clergy. (Such was the case in Ireland before the reform imposed by Paul Cardinal Cullen in the middle of the nineteenth century.) Much of the time of the early leaders of the church was spent trying to patch up quarrels between dissidents and heal minor schisms. Almost all the great leaders of the last century insisted on the need for democratization. Carroll, the first bishop, would not accept the appointment until he was elected to it by his colleagues, and he and his successors repeatedly argued for a return to the ancient practice of election of bishops by clergy and people, insisting that that was the only method of ecclesiastical nomination which would be acceptable in a democratic society like the United States. Their successors of today, perhaps realizing that they could not be elected to anything, have resisted steadfastly any attempts to broaden the base of episcopal nomination.

The most serious problem since the beginning of the history of the Catholic church in the United States up to the present has been immigration. Before the nineteenth century there was conflict between the French clergy (who had come to the United States during the French Revolution) and Irish immigrants. Later there would be conflict between Irish clergy and German immigrants, and then between German and Irish clergy and southern European immigrants. The leadership of each era was caught among many different forces—nativist prejudice against everything foreign, Irish nationalist pressure to support revolutionary movements in Ireland, Irish Americanizing pressure to homogenize the church, and immigrant pressure to keep their own separate subcultures alive. Furthermore, as in the case of the famous Cahensley memorial (written by a sincere German layman who feared for the loss of millions of Catholic immigrants and recommended the establishment of nationality dioceses in the United States), they also had to deal with pressure from European groups who thought they knew the needs of the American church better than the men on the scene. And at all times there was Rome, usually (though not always) trying to be sympathetic, but almost never able to achieve understanding (even perhaps to the present time.)

The "official" histories present the "Americanizers" as the "good guys," and those like Cahensley, who wanted to set up separate national churches for the ethnic groups, as the "bad guys." However, revisionist writers such as Coleman Barry took a more benign view of the efforts of the more recent immigrants to maintain some independence, and argued that on many occasions the Irish were unduly assimilationist and equated being an Irish American Catholic with being an American Catholic.[2] More recently, the eastern and southern European groups have begun to tell their side of the story.*

A somewhat different view of the role of the Irish ecclesiastical leadership is to be found in the recent study by Charles Shannabruch of the response of the Archdiocese of Chicago to immigration.[3] Combining the official and the revisionist perspectives, Shannabruch argues that the Irish bishops of Chicago (and George William Mundelein, who was Irish in style and outlook if not in background) were essentially power brokers, responding with remarkable dexterity to the conflicting pressures of the various forces and strains resulting from massive immigration and vigorous nativist opposition.

Finally, the critical event in American Catholic institutional history was the Roman solution to the multiple controversies of the 1890s. In the encyclical letter *Testem Benevolentiae*, the pope, influenced by the first apostolic delegate (Archbishop Satoli), condemned the heresy of "Americanism" and dealt a death blow to the creative liberalism of American Catholicism. In retrospect it appears that Leo XIII was badly misinformed about the American situation, and that the heresy of American-

---

* See Daniel Buczek, *Immigrant Pastor* (Waterbury, Conn.: Hemingway Corp., 1974), for a description of the life of Monsignor Lucyan Bojnowski, a Polish pastor in New Britain, Connecticut.

ism (to the extent that it existed at all) was to be found in the writings of some French authors.*

Americanism, as explicitly condemned by the encyclical, did not exist in the United States; but it is also true that many of the themes and practices of American Catholic life at the turn of the century, particularly in those dioceses where liberal leadership such as Gibbons, Ireland, Keane, and Spalding were in power, were profoundly threatening to Rome. Ironically, most of these emphases—openness to non-Catholics, strong social concern, consultation with clergy and the laity in decision making, optimism about the modern world, willingness to conduct a dialogue with anyone, endorsement of scientific and technical progress—became church policy after the Second Vatican Council. Phantom heresy or not, Americanism has now become official church policy.

But the condemnation in 1895 was a savage blow to the pride of the young ecclesiastical institution. Now, some eighty years later, it still has not fully recovered—in part because the apostolic delegates who came after the encyclical (up to but not including the present incumbent, Archbishop Jean Jadot) have consistently supported the appointment of "safe" bishops who lack the initiative and the ingenuity of the nineteenth-century pioneer-intellectual bishops. From 1895 on, the American Catholic hierarchy slipped deeper and deeper into mediocrity (always with some notable exceptions), until in the years after the Vatican Council, when it had to deal with the most serious crisis in American Catholic history, the top-level leadership of the American hierarchy was a group of bumbling, incoherent nonentities who were perfectly prepared to do to the church what all the nativist enemies of the immigrant past could not do.

While the ecclesiastical institution was struggling with physical hardship and danger, with the construction of schools, hospitals, and churches, with internal conflicts and external pressures, and, ultimately, with its own incompetence, the immigrants continued to come. There were perhaps 35,000 Catholics in the United States at the time of the Revolutionary War—less than 1 percent of the population; two hundred years later there are fifty million Catholics—25 percent of the population. It is now estimated that about half of those who live in the United States today are descendants of immigrants who came after the Treaty of Paris, and half of those immigrants were Catholic.

---

* Particularly one Abbe Kline, a French writer who commented on a biography of the founder of the Paulist order (*The Life of Father Hecker* by Walter Elliot) and saw in the life and work of Hecker a new theory of church life which he admiringly dubbed "Americanism." American church conservatives and some of the leaders of the immigrant groups used the French controversy over Kline's writings to obtain a Roman condemnation that could have been interpreted as applying to the liberal leadership (many of whom were also assimilationists).

Until much more basic research is done, very little can be said about the social history of Catholics in the nineteenth century. The United States census does not ask a religious question, and the statistics of the Catholic church are incredibly poor.* We know that the immigrants came, we know that most of them were very poor, we know from the research of Oscar Handlin that conditions under which they lived were terrible (though perhaps not as bad as those they left behind).[4] We know that they encountered considerable nativist opposition, and we also know that quite early they became politically active (Irish mayors were elected in the second half of the nineteenth century in many major cities). We know that they began to build schools because the public schools were seen (quite correctly) as Protestant schools bent on converting children away from Catholicism. We know that they worked on canals and railroads, and later in such public occupations as the police department, the fire department, and municipal transportation. We know that many of their daughters became school teachers; we have the myth (caricaturized in the film *The Godfather*) of rapid upward mobility for some—the poor but hard-working immigrant, his successful but conflicted son, and the well-to-do suburban grandson who does not appreciate and is not interested in the struggles of the past. Data to confirm or deny that myth are somewhat conflicting. Stephan Thernstrom presents data that suggest very slow upward mobility for the Irish in Boston, for example, while Daniel P. Moynihan has data that show that the second-generation Irish quite rapidly became members of the middle class.[5]

Estimates of the number of Catholics at any given time are at best guesses. By 1820 there were perhaps 250,000 Catholics; a decade later the number was closer to three-quarters of a million; by 1850 it is estimated that there were close to two million Catholics, and in 1860, almost four million. After that the waves of immigrants came so rapidly that attempts to estimate their numbers became virtually impossible. Immigration records themselves are inadequate, because only Italy was send-

---

* Most of the figures given in the annual *Catholic Directory* are completely unreliable. They are arrived at by adding up questionnaires filled out in each parish house and chancery house in the country. Frequently a pastor will not even leave his desk to check his parish records as he fills out the questionnaire, or will simply shout down the corridor to another priest, "Hey, how many baptisms did we have last year?" To which the priest will shout back a rough guess. In chancery offices bishops frequently instruct their secretaries to "Put down the same thing we put down last year." Furthermore, there is some reason for pastors to want to hide the number of people they have for fear their assessments in the annual collections will go up if the chancery office knows the full truth about the size of their parish (marvelous approach to planning!). Compared with the national survey estimates, the official directory is somewhere between eight and ten million short of the real number of American Catholics in the United States. The Catholic church has no central statistical office, and is not likely to have one, because if you are a bishop you do not need statistics; you get your evidence directly from the Holy Spirit.

ing immigrants who were overwhelmingly Catholic. Both Protestants and Catholics came from Ireland, both Jews and Catholics came from eastern Europe—generally from places which were part of Austria or Russia, since the eastern European Slavic countries had yet to gain independence.*

At one time there was a considerable controversy among American Catholics about the "loss of faith" among immigrants. Early bishops like John England wanted to put pressure on Rome for greater support and reported millions of apostates (in England's case, three and three-quarter million). Later, Peter Paul Cahensley would double that figure to warn Rome of the dangers of Irish incompetence in the administration of the American church. In 1925 Gerald Shaughnessy put the defections at 225,000, and those mostly before 1820.[6] Contemporary scholars think that that number is too high.[7] Surely there was almost no apostasy among the German immigrants, and the survey data show very little among the Italian and the eastern European immigrants and their descendants. Thus if there was massive apostasy, it would have occurred among Irish immigrants before, as Shaughnessy suggests, the American church became well organized. There were some losses, especially among those Irish who migrated to the southern part of the United States, where there were no priests and no organized church. But a substantial number of the immigrants with Irish names who migrated to the South were apparently Protestant before they came to the United States. There is no way of knowing how many of those Americans who describe themselves today as Irish and Protestant (a larger number, by the way, than those who describe themselves as Irish and Catholic) are descendants of those who came to the United States as Catholic but whose family left the church sometime after immigration. All things considered, however, one must conclude with Shaughnessy that any apostasy problem was probably resolved after 1820. Apostasy would become a major problem for American Catholics only in the 1970s, after the publication of the encyclical letter *Humanae Vitae*—long after the immigrant era had come to an end.

But almost everything that is said about the American Catholic population in the nineteenth century is guesswork, even though some of it is slowly being documented by careful historical research. When we enter the twentieth century, however, it becomes possible to reconstruct the social profile of Catholics from the early 1900s to the present. There are still alive today, and available to national survey interviewers, observers who can report on the state of things at the turn of the century. In fact, some of them automatically fall into national survey samples.

* Thus providing some American Catholics of Slavic background with an opportunity to claim Austrian rather than Slovak, Slovene, or Croatian backgrounds when they had made a little money and moved into expensive neighborhoods.

Ideally, one should screen for a large sample of Catholics in their
seventies and eighties and ask them a detailed series of questions about
the educational, economic, political, and occupational condition of their
family during childhood and adolescence, as well as some questions
about the experiences of their parents. This might enable us to move back
another quarter of a century. Such a project of historical reconstruction
would be difficult and expensive(and no one is likely to fund it), but it
would solve quickly and decisively the problem of the absence of data
on the condition of immigrant groups in the late years of the nineteenth
and early years of the twentieth centuries.

Short of such heroic measures, one can still trace some of the social
and demographic history of Catholicism in the twentieth century from
existing survey data.* People who were in their seventies during the
last decade, born just before or just after the turn of the century, can
give us certain information about themselves and their families that will
give us some picture of what was going on in the Catholic collectivity at
that time. We know, for example, that in 1974, 85 percent of American
Catholics were born in this country, that 40 percent of them had at least
one foreign-born parent, and that 80 percent had at least one foreign-
born grandparent. One can then say that at the present time 15 percent
of the American Catholic population is first generation, 25 percent is
second generation, 40 percent is third generation, and 20 percent is
fourth generation—two-fifths of American Catholics are either immigrants or the children of immigrants. If one cross-tabulates this generational background by age cohort, one can begin to construct a
generational distribution of American Catholics back into the early years
of this century—and indeed, we can do so for each of the major ethnic
subcollectivities within the larger Catholic population. (See table 2.1)

What is striking about the data in Table 2.1 is how close American
Catholics are to the immigrant experience. Most Americans assume that
with the passing of restrictive immigration laws in 1920, the immigration
era came to an end (even though thousands of immigrants still come to
the United States each year). But the end of an era does·not mean the end
of that era's impact. Some of the people whose formative experiences
took place during that era are still very much alive, and their children
constitute a large proportion of the Catholic population; indeed, the
first generation are still the majority of the eastern and southern European Catholic groups and the majority of all Catholics over forty. One
simply cannot talk sensibly about American Catholics without realizing
that, even in the middle 1970s, they are still an immigrant people. Yet

---

* We now begin to use "SuperNORC," a composite sample made up of
twelve national sample surveys collected by NORC during the late 1960s and 1970s.

TABLE 2.1

First and Second Generation by Ethnic Group
(percent)

ETHNIC GROUP	FOREIGN BORN OR NATIVE BORN OF FOREIGN PARENTS
Irish	$26_{(536)}$
German	$24_{(642)}$
Polish	$52_{(333)}$
Slavic	$63_{(383)}$
Italian	$72_{(689)}$
French	$39_{(229)}$

virtually all discussion of the intellectual and economic achievement of American Catholics, as well as the crisis of change in the Catholic church, ignores the overwhelming fact of immigration. The proportion of first or second generation in America varies greatly according to the ethnic group. Thus in NORC's large composite sample (for those surveys where generation was asked) about a quarter of the Irish and Germans were foreign born or had foreign-born parents, but 52 percent of the Poles, 63 percent of the Slavs, and 72 percent of the Italians were either first (foreign born) or second (native of foreign-born parents) generation. The French were between the early and later immigrants, with about two-fifths being first or second generation (most of them presumably in New England and not Louisiana or Texas—the two principal locations of French settlement).

What can we say about the experience and achievements of American Catholics since the turn of the century? To what extent have they become members of the middle class?

The problem is that both the definitions and size of the middle class have increased dramatically over the years. In 1920 you could be a respectable member of the middle class without going to college; in 1976 it would be pretty hard to be so. But despite higher "admission standards" to the middle class, it still has far more members because the proportion of young Americans (including Catholics) going to college has more than doubled since the beginning of the century. If most Catholics did not belong to the middle class at the turn of the century, neither did most Americans; if a very large number of them belong to the middle class today, so do most Americans. So the term "middle class" (or "white collar") is too imprecise for careful analysis, but one can turn to a much more specific question: What proportion of Catholics attend college, and how does that proportion compare with both the national average for college attendance and the college attendance of other denominational groups?

College attendance is something that happens nominally at one specific point in the life cycle. Despite the occasional older person who returns to complete a college education, most people attend college between the years of seventeen and twenty-five. Thus if we know the proportion of a given population within a specific age group that attended college, we are able to learn some important things about the state of that population at the time when the age cohort was in its college years. We can estimate how important college attendance seemed to be for the group, how willing young people were to defer marriage and career to complete their educations, what kinds of ambitions for upward mobility were at work in the group, and how willing the members were to risk time and resources in what might have seemed a risky venture. The proportion of Catholics in their seventies who attended college tells us something important about the state of Catholic collectivity sixty years ago.

In the Edwardian era, the decade just before World War I, 17 percent of the American population went to college, as did 7 percent of the Catholics (table 2.2). During that turn-of-the-century era, Catholics were the least likely of all American denominational groups to attend college. The Catholic population was busy absorbing the tremendous weight of the new eastern and southern European immigrants. However, in the next two cohorts, those who grew up during World War I and the Roaring 20s, young Catholics moved even with Lutherans and ahead of Baptists in their college attendance; and during the Depression and World War II, the new cohorts coming of age managed to maintain their lead over the Baptists, pull ahead of the Lutherans, and, by World War II, come to within one percentage point of the national average. Finally, in the Vietnam era, Catholics became slightly more likely than the American average to go to college; they pulled even with the Methodists, and their rate of college attendance was surpassed only by the Jews, the Presbyterians, and the Episcopalians. In this same seven-decade period, Jews, who by the turn of the century had already achieved the national average in college attendance, quickly moved into second place behind Episcopalians and Presbyterians; then, in the 1930s and 1940s, Jews moved into their present position as the best educated of all American denominational groups. To put the matter somewhat differently, the Catholic Vietnam generation achieved the same level of college attendance as the Jewish Depression cohort.

Figure 2.1 shows the logarithmic slopes of the odds ratio of the proportion going to college divided by the proportion not going to college (for Catholics, Jews, and Methodists—assuming that the last group is the "typical" American Protestant denomination). The Jewish slope

TABLE 2.2

Proportion of Denominational Groups Attending College by Cohort—Non-Spanish-Speaking Whites Only
(percent)

DENOMINATIONAL GROUP	EDWARDIAN	WORLD WAR I	ROARING '20's	DEPRESSION	WORLD WAR II	COLD WAR	VIETNAM
Baptists	10(71)	8(178)	11(256)	20(312)	22(294)	23(374)	28(258)
Methodists	22(92)	25(184)	26(249)	29(294)	35(284)	36(244)	45(170)
Lutherans	13(47)	8(107)	15(164)	17(203)	27(206)	30(220)	43(150)
Episcopalians	32(23)	32(36)	45(49)	44(69)	45(47)	54(59)	65(34)
Presbyterians	48(54)	56(176)	53(112)	57(131)	47(108)	71(92)	65(74)
Catholics	7(100)	15(235)	14(452)	19(610)	29(715)	31(670)	45(532)
Jews	17(18)	29(24)	42(53)	47(76)	69(67)	64(63)	88(49)
All	17	18	18	23	29	32	43

SOURCE: Andrew M. Greeley, *Ethnicity, Denomination, and Inequality*, p. 32. (This and following tables and figures from this source are reprinted from *Sage Research Papers in the Social Sciences*, vol. 90-029, by permission of the publisher, Sage Publications, Inc. — © 1976.)

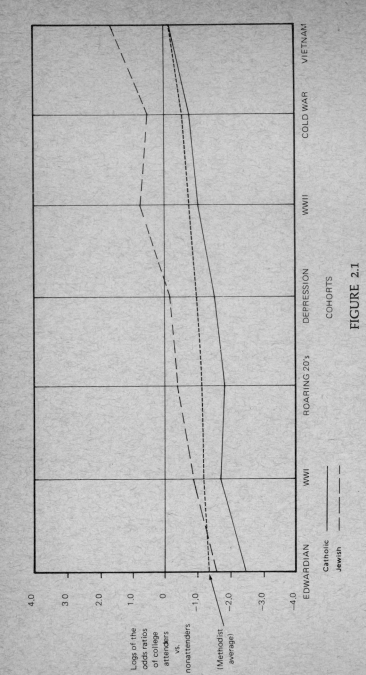

FIGURE 2.1

Catholic and Jewish college attendance over time.

SOURCE: Adapted from Greeley, *Ethnicity, Denomination, and Inequality,* pp. 34–35.

climbs sharply, crossing the Methodist slope during the World War I cohort; the Catholic slope climbs more gradually, intersecting with the Methodist slope only during the Vietnam era. The Jews caught up before the first quarter of a century was over; it took the Catholics three-quarters of a century to catch up. (All three slopes, incidentally, are significantly different from each other, thus indicating that the different shape of the slope is not due to chance variation.)

So it is only in the era after World War II, particularly the last two decades, that the Catholic propensity to go to college began to match that of typical American Protestants. The picture that emerges is of several decades in which the Catholic collectivity was overwhelmingly first- and second-generation immigrant while slowly gaining on the rest of the population on its capacity to send its children to college. The relative growth of Catholic college attendance was not appreciably impeded by the Depression, but the big breakthrough came with the prosperity in the years after World War II. However, the relative upward mobility of Catholics in the last three decades (compared with Protestants) is not merely a function of the increase in higher education in the whole population. The Catholic slope is sharper than the Methodist group; both groups were going up as a result of the ideal of universal higher education, but Catholics were moving up more rapidly.

The different Catholic ethnic groups, however, had varying stories to tell in the years since the turn of the century (Table 2.3). The Irish had already surpassed the national norm in the proportion going to college by the end of the second decade of the twentieth century (the Edwardian era is omitted because there were not enough respondents in most of the ethnic groups for that period). They would maintain their relative position above the national average for the remaining decades until the present, slowly creeping up and finally passing British Protestants in their propensity to go to college. The Irish in the Vietnam generation were more likely to go to college than any other white Gentile group in American society (though they tied with the Scandinavians).* Except for a sharp dip during the Roaring '20s, German Catholics have stayed close to the national average in their propensity to go to college. It may well be that the sharp drop in the 1920s was a result of the negative effects of the anti-German feeling with which young people grew up in the wake of World War I. However, if such be the case, the same effect is not observable among German Protestants.

For the Polish, Slavic, and Italian ethnic groups, however, the story is very different (Figure 2.2). From the decades after World War I until the Depression, these groups were at the bottom of the college attendance scale, and their slope moved up only very slightly. But from World

* The "American" Protestants are those respondents who do not have a specific ethnic background or who cannot choose among several different ones.

TABLE 2.3

Percent Attended College by Cohort by Ethnic Group

ETHNIC GROUP	WORLD WAR I	ROARING '20's	DEPRESSION	WORLD WAR II	COLD WAR	VIETNAM
Protestant						
British	29 (221)	34 (310)	38 (366)	42 (334)	43 (306)	54 (183)
German	16 (198)	20 (156)	23 (310)	28 (329)	33 (324)	38 (224)
Scandinavian	16 (64)	21 (85)	20 (86)	40 (89)	32 (98)	59 (65)
Irish	16 (90)	18 (136)	26 (138)	25 (132)	28 (131)	34 (85)
"American"	13 (286)	17 (378)	24 (481)	25 (521)	29 (521)	34 (349)
Catholic						
Irish	24 (56)	26 (147)	32 (200)	43 (167)	38 (167)	59 (101)
German	24 (40)	12 (165)	19 (218)	29 (186)	32 (186)	45 (97)
Polish	7 (15)	4 (78)	9 (160)	20 (82)	34 (82)	49 (61)
Slavic	9 (22)	11 (96)	12 (176)	21 (74)	36 (74)	42 (55)
Italian	7 (55)	9 (139)	14 (261)	21 (171)	29 (171)	45 (125)
French	11 (18)	13 (63)	19 (85)	21 (108)	22 (183)	19 (41)
All	17	18	23	29	32	43

Total = 17,222

SOURCE: Greeley, *Ethnicity, Denomination, and Inequality,* p. 62.

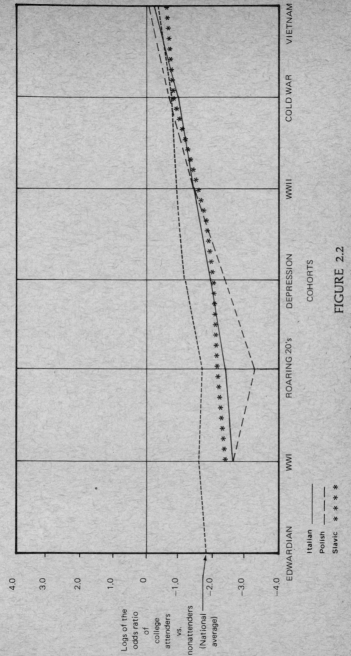

FIGURE 2.2

Italian, Polish, and Slavic college attendance over time.

SOURCE: Adapted from Greeley, *Ethnicity, Denomination, and Inequality*, pp. 65–67.

War II to the present, college attendance for the three eastern and southern European Catholic groups moved up very sharply, coming virtually even with the national average during the Cold War decade and, in the case of the Italians and Poles, actually crossing it during the Vietnam years.

The relative improvement of American Catholics in college attendance through the last seven decades, and in particular the three after World War II, is divided into two parts. The Irish and the German Catholics were either above or at the national average by 1910; the Poles, Italians, and other eastern European Catholics moved up sharply after 1940. In other words, if it had not been for the influx of southern and eastern European migrants, Catholics would have achieved educational parity by the turn of the century. (A complex reconstruction through investigation of parental education indicates that by the beginning of the Edwardian era—1900—the Irish were already at the national average in college attendance.) Roughly speaking, it took until 1940 for the Catholic collectivity to absorb the southern and eastern European immigrant groups to such an extent that they could begin to send their young people to college in the same proportion as that of other white Americans. If the capacity to send one's children to college in due proportion is a measure of acculturation, and if one realizes that massive immigration of southern and eastern European Catholics ended only in 1920, one would conclude that the acculturation of the eastern and southern European groups took a quarter of a century or less. The achievement was not as dramatic as that of the Jewish immigrants who came at the same time, who seemed to be acculturated on the spot, but it was still a rather impressive performance for a group whom the Dillingham Immigration Commission concluded, as part of their recommendation of restrictive legislation, could not easily be absorbed into American society. Whether the eastern and southern European groups were absorbed or whether they set about vigorously to absorb themselves may be a matter for debate, but whatever it was happened much more quickly than most people thought—perhaps more quickly than most people realize even today.

The picture, which is part of American mythology, of the children and grandchildren of immigrants struggling upward in American society is generally validated by the research evidence available to us, but there are three notable surprises: (1) the Irish and the Germans had achieved success in America (if the ability to send your children to college is considered a mark of success) by the turn of the century; (2) after the Depression was over, the eastern and southern European groups began a climb upward that was not as dramatic as the upward surge of the Jews but was still quite unexpected (and to a considerable extent still unacknowledged); (3) at the present time, Irish Catholics are the most likely of any American white Gentile ethnic group to send their children to college, and the Polish, Italian, and Slavic groups have now surpassed the national average in college attendance.

In Chapter 4 we will see what Catholic ethnic groups have done with their educational achievement, but if parity with the national performance is a mark of acculturation, then American Catholics are now thoroughly acculturated to American society. The Irish and the Germans were acculturated by the turn of the century, and the eastern and southern European groups achieved their parity in the years since the end of World War II. The myth of the blue-collar ethnic has not been completely eliminated, but it now must be called into serious question. A dramatic change has taken place among the American Catholic population since the 1940s, one that scarcely anybody, including Catholics themselves, has noticed.

REFERENCES

1. The classic works are John Tracy Ellis, *American Catholicism* (Chicago: University of Chicago Press, 1956); Thomas T. McAvoy, *History of The Catholic Church in America* (Notre Dame, Ind.: University of Notre Dame Press, 1969); Theodore Maynard, *The Story of American Catholicism* (New York: Macmillan, 1943; Doubleday Image paperback, 2 vols., 1960). Not recommended is the shallow study by John Cogley, *Catholic America* (New York: Dial Press, 1973).

Peter Guilday's *The Life and Times of John Carroll, Archbishop of Baltimore* (New York: Encyclopedia Press, 1922), and *The Life and Times of John England, First Bishop of Charleston* (New York: American Press, 1925), are vast, tedious, and comprehensive biographies of the old style. Far more readable is John Tracy Ellis' *Life of James Cardinal Gibbons*, 2 vols. (Milwaukee: Bruce, 1952). Among better, more recent works are Kathleen Healy, *Frances Ward, American Founder of the Sisters of Mercy* (New York: Seabury Press, 1973); Thomas O'Brien Hanley, *Charles Carrolton, the Making of a Revolutionary Gentleman* (Washington, D.C.: Catholic University of America Press, 1970); and three volumes by Coleman J. Barry: *Worship and Work* (the history of St. John's Abbey and University) (Collegeville, Minn.: Liturgical Press, 1956); *American Nuncio* (Collegeville, Minn.: St. John's University Press, 1969); and *Upon These Rocks: Catholics in the Bahamas* (Collegeville, Minn.: St. John's Abbey Press, 1973).

A superb account of the critical turn-of-the-century period by a non-Catholic writer is Robert D. Cross, *The Emergence of Liberal Catholicism in America* (Cambridge, Mass.: Harvard University Press, 1958). Two interesting volumes of memoirs are George N. Schuster, *The Ground I Walked On* (New York: Farrar-Straus, 1961), and Michael Harrington, *Fragments of the Century* (New York: Saturday Review Press, 1973).

In the sociological area, *Priest and People* by Joseph Fichter (New York: Sheed & Ward, 1965) is embarrassingly bad. So too are his *America's Forgotten Priests: What Are They Saying* (New York: Harper & Row, 1968) and *The Catholic Cult of the Paraclete* (New York: Sheed & Ward, 1975). In *Catholics in College: Religious Commitment and the Intellectual Life* (Chicago: University of Chicago Press, 1967), Professor James W. Trent displays a lamentable inability to understand the mechanics of drawing a national sample or the importance of representativeness in such a sample. John L. Thomas, *Religion and the American People* (Westminster, Md.: Newman Press, 1963), is a report on the first national sample survey of Catholics. Unfortunately it was written after the data cards were destroyed and hence deals only with cross-tabular comparisons of Protestants, Catholics, and Jews. A report on a later follow-up survey is contained in Martin Marty, Andrew M. Greeley, and Stuart E. Rosenberg, *What Do We Believe?* (New York: Meredith, 1968). Gerhard Lenski, *The Religious Factor*, (Garden City, N.Y.: Doubleday Anchor, 1961) will be discussed at length in Chapter 4; in its sweeping generalizations based on a small sample of Catholics in one city, Lenski's book comes dangerously close to bigotry.

Three collections of essays, some of them allegedly sociological, have been gathered by members of the Notre Dame University faculty: Lewis J. Putz, ed., *Catholic Church U.S.A.* (Notre Dame, Ind.: University of Notre Dame Press, 1956); Leo R. Ward, ed., *Catholic Life U.S.A.* (New York: Herder and Herder, 1959); and Philip Gleason, ed., *Contemporary Catholicism in the United States* (Notre Dame, Ind.: University of Notre Dame Press, 1969). In the last-named volume only the essays by Gleason and James Vanecko are worth reading.

There is a relatively extensive literature on Catholic social involvement: David O'Brien, *American Catholics and Social Reform* (New York: Oxford University Press, 1968); George Hugh Flynn, *American Catholics and the Roosevelt Presidency, 1932–1936* (Lexington, Ky.: University of Kentucky Press, 1968); Aaron I. Abell, *American Catholicism and Social Action* (New York: Hanover House, 1960); Philip Gleason, *The Conservative Reformers: German-American Catholics in the Social Order* (Notre Dame, Ind.: University of Notre Dame Press, 1968); and Mary Harrita Fox, *Peter G. Dietz, Labor Priest* (Notre Dame, Ind.: University of Notre Dame Press, 1953).

An interesting collection of observations by foreign visitors is Dan Herr and Joel Wells, *Through Other Eyes: Some Impressions of American Catholicism by Foreign Visitors from 1777 to the Present* (Westminster, Md.: Newman Press, 1964). Most of the foreign visitors, alas, did not know what they were talking about. Nor is there much insight in Philip Scharper, ed., *American Catholics, a Protestant-Jewish View* (New York: Sheed & Ward, 1959).

Some very bad books have been written about the church in the years after the Second Vatican Council: Michael Novak, *The Open Church* (New York: Macmillan, 1964); Daniel Callahan, *The Mind of the Catholic Layman* (New York: Scribner's, 1963) (by "the" Catholic layman, Mr. Callahan means himself); Thomas F. O'Dea, *The Catholic Crisis* (Boston: Beacon, 1968); and, from the more conservative viewpoint, James Hitchcock, *The Decline and Fall of Radical Catholicism* (New York: Herder & Herder, 1971).

The best of the contemporary professional historians writing about Catholicism are much less concerned with institution than they are with people. They are free of the pieties of some of the earlier writers. Some examples: Charles Shannabruch, "The Chicago Catholic Church's Role as Americanizer, 1893–1928" (Ph.D. diss., University of Chicago, 1976); Jay P. Dolan, *The Immigrant Church: New York's Irish and German Catholics, 1815–1865* (Baltimore, Md.: Johns Hopkins University Press, 1975); Paul R. Messbarger, *Varieties of Accommodation: Social Uses of American Catholic Literature, 1884–1900* (Boston: Boston University Press, 1971); and my own attempt at a sociological interpretation of the history of American Catholicism, *The Catholic Experience* (Garden City, N.Y.: Doubleday Image, 1969).

2. Coleman Barry, *The Catholic Church and German Americans* (Milwaukee: Bruce, 1953).

3. Shannabruch, "Chicago Catholic Church's Role."

4. Oscar Handlin, *The Uprooted* (Boston: Little, Brown, 1951).

5. Stephan Thernstrom, *The Other Bostonians: Poverty and Progress in the American Metropolis, 1880–1970*, Harvard Studies in Urban History Series (Cam-

bridge, Mass.: Harvard University Press, 1973); Nathan Glazer and Daniel Patrick Moynihan, ed., *Ethnicity: Theory and Experience* (Cambridge, Mass.: Harvard University Press, 1975).

6. Gerald Shaughnessy, *Has the Immigrant Kept the Faith?*, Religion in America Series (1925; reprinted, New York: Arno Press, 1969).

7. See, for example, Victor Greene, *For God and Country: The Rise of Polish and Lithuanian Ethnic Consciousness in America* (Madison, Wis.: The State Historical Society of Wisconsin, 1971).

CHAPTER
3

Economic and
Educational Achievement

We NOW TURN from attempts to reconstruct the past to a portrait of the present-day educational, occupational, and economic status of American Catholics. Two social science terms must be defined at the beginning—"stratification" and "mobility." To social scientists "stratification" means where you are on one of the standard measures of "social class"—education, occupation, income. "Mobility" means where you are, given where you have come from. There is a good deal of theoretical wrestling within the discipline about more precise meanings of the words, and a lot of theoretical agonizing because terms like "class" and "status" go back to the very beginnings of modern social theory. For the purposes of this portrait of American Catholics, however, our definitions will be operational of necessity, because our data are limited to a rather small number of variables. The stratification of American Catholics will be measured by education, occupation, and income. Their mobility will be measured by their educational attainment, given the educational attainment of their parents, their occupational prestige, given their own and their parents' education, and their income, given their own and parental education and their occupational prestige level. We will ask how much education Catholics get, compared with members of other denominational groups who come from families whose parents had the same amount of education. Then we will ask how well Catholics do occupationally when compared with those whose education and parental education is similar. Finally, we will ask how much money Catholics make when they are compared with others who have the same jobs, the same parental education, and the same personal educational attainment.

This kind of analysis has been carried out admirably by such sociological greats as Otis Dudley Duncan, James Coleman, Peter Blau, and William Sewell.[1] Ideally one should know the occupation of the father of the respondent and as much other pertinent information as one can gather. However, the constraints of data available to us are such that we simply do not have that information, and we must make do with five variables—mother's education, father's education, respondent's education, respondent's occupational prestige, and respondent's income.

There has been a vast amount of debate about the economic achievement of Catholics in America.[2] The best articles on the subject, by Bogue and Gockel, do not indicate much difference between Catholics and Protestants in educational, occupational, and economic achievement. But much of the other literature, especially Gerhard Lenski's "The Religious Factor," argued that Catholics did poorly compared with others in American society partly because their religious principles frowned on worldly success, partly because rigid family controls inhibited the development of the virtues necessary for economic achievement, partly because of the responsibilities of raising larger families, and partly because the "spirit of Catholicism" was really foreign to a technological world. Lenski thought he found evidence of this in his Detroit area study. Howard Schuman, in a later replication of Lenski's research, could find no such evidence.[3] Albert J. Mayer and Harry Sharp, in another analysis of the Detroit area study, also reported economic deficiencies for Catholics: "The powerfully reinforced and traditional Roman Catholic church tends to orient its members toward the hereafter; successful performance in the marketplace and the acquisition of symbols of economic achievement are of relatively little importance as an indication of the Catholic's status after death."[4]

David McClelland reported on the administration of his need-achievement test to a national sample in both Germany and the United States and found a mixed picture.[5] Some Catholics had high need-achievement and others did not. His associates, Veroff, Feld, and Gurin, administered the need-achievement test to a national sample of respondents and discovered that 48 percent of the Protestants scored high on the presumed measure of the Protestant ethic, as did 57 percent of the Catholics and 68 percent of the Jews.[6] This was an interesting fillip to the theory of the Protestant ethic; Jews were more Protestant than Catholics, who were more Protestant than Protestants.*

---

* It might be noted in passing that Max Weber himself was dubious about whether the differences between Protestants and Catholics on the "Protestant ethic" (the drive for this-worldly achievement) would survive a long period of industrializa-

Data have been unavailable for a solution to this not always edifying discussion. The 1957 census on religion taken by the Current Population Survey was locked up in a government safe at the instigation of the civil libertarian organizations (a curious position for them), and by the time the Freedom of Information Act forced the release of tabulations from this survey, the data processing cards had vanished. The typical national sample survey does not contain enough respondents to permit detailed analysis of the stratification and mobility of American denominations and ethnic groups. Neither Gockel nor Bogue, the best contributors to the discussion, had large enough samples or ethnic variables. Hence, until the recent National Opinion Research Center (NORC) study, *Ethnicity, Denomination, and Inequality,* based on a composite of twelve NORC samples, there was no means available to resolve the debate, especially since it is unlikely that the federal government will ever permit the Census Bureau to ask a religious question—however important religious differentiation may be in the United States.

National sample surveys generally show Catholics with higher income levels than Protestants, but a typical national sample of 1,500 respondents is too small to permit detailed analysis, and these findings can be too easily dismissed by social scientists as showing the depressing effect on Protestant incomes of the disproportionately black, rural, and southern composition of the Protestant population. Even a sample the size of the NORC composite (just under 18,000) is not big enough to enable us to compare southern rural Catholics with southern rural Protestants, so critical comparisons have to be made in the metropolitan areas of the North ("metropolitan" meaning one of the standard metropolitan statistical areas of the country, and "north" meaning northeast middle Atlantic, east north central, west north central, mountain, and Pacific geographic regions). Spanish-speaking Catholics and blacks are excluded from the analysis because better data on them are available from the census and because one wishes to exclude the cross-cutting influence of race from any attempt to measure stratification and mobility of reli-

---

tion. In attempting to prove that differences between Protestants and Catholics do persist in modern society, Lenski went much further than Weber. Says Lenski: "With considerable regularity, Jews and white Protestants have identified themselves with individualistic, comparative patterns of thought and actions linked with the middle class historically associated with the Protestant ethic. . . . By contrast, Catholics . . . have more often been associated with the collectivistic, security-oriented working-class patterns of thought and action."[7] Not content with such generalization based on tables sometimes consisting of only thirteen Catholics and twenty Protestants, Lenski goes on to say that the critical weakness in Catholicism is "the basic intellectual orientation which Catholicism develops, an orientation which values obedience above intellectual autonomy. Also influential is the Catholic tendency to value family and kin group above other relationships. . . . The implications of this for the future of American society are not difficult to discover."[8] Lenski omits the precise implications, but they are obviously not good.

gious denominations. In addition, a number of the groups to be considered in our analysis are residual. "British" includes Scotch and Welsh as well as English; "Scandinavian" includes Norwegians, Swedes, Danes, and Finns; "Slavic" includes those eastern Catholics who are not Poles; and "American" Protestants are those who have no self-conscious ethnic identification or who could not choose among several ethnic backgrounds. Such fascinating groups as Greeks, Chinese, Japanese, and Indian Americans had to be excluded from the analysis simply because there are not enough respondents in those categories. Similarly, all Protestant denominations smaller than Episcopalians and Presbyterians had to be excluded, and there was no way to differentiate among the various sub-denominations within the Methodist, Lutheran, and Baptist churches.

The "occupational prestige" score, developed by the National Opinion Research Center, is one in which occupations are ranked according to the prestige they are given in American society, a prestige which is closely connected with the education required for the occupation and the income provided by it. Income is measured by the mean family income stated in 1974 dollars.

In order to keep the number of tables at a minimum, I shall present two tables for each of the three measures of stratification—income, occupation, and mobility. One of the tables will deal with denominations, the other with ethnic groups. Each table will be divided into two panels, the first (A) presenting data for the entire country, the second (B) presenting data for metropolitan areas in the North. Each panel will consist of two columns, the first presenting the deviation above or below the average for the given group and constituting our measure for stratification. The second column will present the deviation from the average once all the pertinent background variables have been taken into account, and will serve as a mobility measure. Thus by glancing at panel A of the table the reader will be able to see how Catholics do in both stratification and mobility on the particular measure in the whole country; by glancing at panel B he will be able to see what Catholic achievement is in the urban areas of the North.*

Thus we can observe in Table 3.1A that the Catholic average for years of education is exactly on the national average of 11.5, slightly behind the Methodist and substantially behind Jewish, Episcopalian, and Presbyterian educational attainment—though somewhat ahead of Lutherans and Baptists. The parents of Catholics had less school than the parents of Protestants, with the average years of school for Catholic

---

* Briefly, the technique for arriving at the mobility measure is based on a multiple regression equation in which the β is considered a conversion rate for a given group. Further details about the technique of "standardizing" stratification measures—along with references to articles which provide even greater detail—can be found in Greeley, *Ethnicity, Denomination, and Inequality*.

fathers being 8.6 while the national average was 9.1. If one takes into account the education of parents and asks whether, relatively speaking, Catholics are more mobile than other groups, one can see from the second column in the first panel (Table 3.1A) that, given where they came from educationally, Catholics do rather well. In the mobility measure they are eight-tenths of a year higher than the average—behind the Jews in educational mobility but ahead of everyone else. Nationally, then, it can be said that in terms of educational stratification, American Catholics are on dead center, but in terms of educational mobility, Catholics are second only to Jews.

TABLE 3.1A

Differences from National Average in Educational Achievement of American Denominational Groups—Non-Spanish-Speaking Whites Only (years of education)

DENOMINATIONAL GROUP	GROSS DIFFERENCE FROM MEAN	NET DIFFERENCE FROM MEAN*
Jews	2.5	2.2
Episcopalians	2.0	−0.3
Presbyterians	1.2	0.4
Methodists	0.4	0.1
Catholics	0.0	0.8
Lutherans	−0.3	−0.1
Baptists	−0.8	−0.2
National Average = 11.5		

SOURCE, tables 3.1A and B: Greeley, *Ethnicity, Denomination, and Inequality*, p. 22.

*Taking into account mother's and father's education.

But as is well known, Catholics (and Jews) are disproportionately located in the cities outside of the southern part of the United States. Could there be greater educational opportunities in the northern metropolitan areas which account for higher achievement for Catholics than might have been expected?

We can observe in the second panel (Table 3.1B) that in the metropolitan regions outside the South, the Catholic educational attainment dips to four-tenths of a year beneath the twelve years average for northern metropolitan regions. The relative position of Catholics remains unchanged. Similarly, the educational mobility measure for Catholics drops from eight-tenths of a year to three-tenths of a year. Nonetheless, even in cities in the North Catholics are still the second most mobile group educationally, although, with the exception of the Jews at the top and the Baptists at the bottom, these measures are not very different for any of the remaining groups.

TABLE 3.1B

Differences from Average in Educational Achievement of American
Denominational Groups in Metropolitan Regions Outside
the South—Non-Spanish-Speaking Whites Only
(years of education)

DENOMINATIONAL GROUP	GROSS DIFFERENCE FROM MEAN	NET DIFFERENCE FROM MEAN*
Jews(315)	2.0	1.8
Episcopalians(185)	1.7	0.2
Presbyterians(325)	0.9	−0.1
Methodists(513)	0.4	0.1
Catholics(3898)	−0.4	0.3
Lutherans(522)	−0.5	−0.2
Baptists(353)	−0.9	−0.8
National average for metropolitan regions outside the South = 12.0		

*Taking into account mother's and father's education.

Thus the upward climb of the Catholic ethnic has produced a situation in which Catholics have caught up educationally with the rest of the country for all practical purposes.

They are also at dead center in occupational prestige (Table 3.2). Catholics are on the average one-tenth of a unit above the national average of 40.3 in occupational prestige, slightly behind the Methodists, substantially behind the elite religious groups (Jews, Episcopalians, and Presbyterians), but ahead of Lutherans and Baptists (Table 3.2A). Similarly, their occupational mobility is one-tenth of a unit above the average (mobility based on the educational prestige Catholics achieved, given their own and their parents' education).

However, it seems safe to assume that there are more prestigious occupations in the cities of the North, and when that fact is taken into account, the occupational prestige of Catholics slips 1.3 units beneath the northern average of 41.8, and the Catholic mobility measure falls to −0.6 (Table 3.2B). These are relatively minor differences and do not affect the Catholic position in the mobility ordering. Nonetheless, the differences do raise a question. Why are Catholics able to make effective use of their parents' education to achieve disproportionately in the educational world, and then are unable to do appreciably better than average occupationally with the education they have received? There are a number of possible explanations. It may be that Catholic culture produces educational ambition which has enabled Catholics to move dramatically beyond the education of their parents but does not provide them with the further ambition and need-achievement that is required for occupa-

TABLE 3.2A

*Differences from National Average in Occupational Prestige of American
Denominational Groups—Non-Spanish-Speaking Whites Only
(0-99)*

DENOMINATIONAL GROUP	GROSS DIFFERENCES	NET DIFFERENCES*
Jews	8.50	7.00
Episcopalians	7.90	4.10
Presbyterians	6.00	3.30
Methodists	0.80	0.10
Catholics	0.10	0.10
Lutherans	−1.40	−1.20
Baptists	−3.10	−2.00
National average = 40.3		

SOURCE, tables 3.2A and B: Greeley, *Ethnicity, Denomination, and Inequality,*
pp. 25, 26.

*Taking into account parental and respondent's education.

tional success. Catholics do "all right" occupationally, but they do not
have the ambition, industry, or need for achievement to carry them
toward greater occupational success. Perhaps their large families or relatively closed neighborhoods explain in part an occupational mobility
which is less impressive than their educational mobility.

It may also be that Catholics feel diffident when facing a non-
Catholic world, or perhaps there may be some discrimination which
prohibits Catholic admission to the higher levels of professional and
occupational success. In any event, the standard deviation of the Catholic
occupational prestige score (15.1) is almost three units less than the
standard deviation of Episcopalians, Presbyterians, and Jews. This means
that Catholics tend to be bunched much more closely around their aver-

TABLE 3.2B

*Differences from Average in Occupational Prestige of American
Denominational Groups in Metropolitan Regions Outside the
South—Non-Spanish-Speaking Whites Only*

DENOMINATIONAL GROUP	GROSS DIFFERENCES	NET DIFFERENCES*
Jews	6.6	5.5
Episcopalians	7.7	6.2
Presbyterians	5.5	3.6
Methodists	1.3	0.7
Catholics	−1.3	−0.6
Lutherans	−2.1	−1.7
Baptists	−5.1	−3.8
Average in metropolitan regions outside the South = 41.8		

*Taking into account parental and respondent's education.

age score than the three elite groups. Such a bunching merely proves that Catholics are less likely than the other three groups to be among the higher levels of occupational prestige. It does not say why they are not there.

One must observe, however, that if such a finding was reported for any minority nonwhite group in American society, the assumption would be automatically that the group in question was a victim of discrimination.

TABLE 3.3A

Differences from National Family Income Average of American Denominational Groups (1974 Dollars)—Non-Spanish-Speaking Whites Only

DENOMINATIONAL GROUP	GROSS DIFFERENCES	NET DIFFERENCES*
Jews	$ 3,387	$ 1,460
Catholics	1,421	1,656
Episcopalians	1,079	−199
Presbyterians	1,023	−988
Methodists	210	− 42
Lutherans	− 555	−123
Baptists	−1,260	−473
National average = $9,953		

SOURCE, tables 3.3A and B: Greeley, *Ethnicity, Denomination, and Inequality*, p. 28

*Taking into account parental education, respondent's education, and respondent's occupational prestige.

The income achievement of Catholics (Table 3.3A) provides little support for a theory of Catholic lack of industry or ambition. Catholics earn more money than any other Christian denomination—$1,421 above the national average and $1,211 above the northern metropolitan average. In both cases, Catholics have a firm grip on second place in the American family income sweepstakes. They are more than $300 ahead of Episcopalians nationally and almost $700 ahead of the third-place Methodists in northern cities. Furthermore, when one turns to the second column, one can observe that on the mobility measure Catholics make more money than Jews when compared with those who have the same educational and occupational background. On the mobility measure nationally, Catholics are almost $200 ahead of Jews, and in the metropolitan regions of the North they are more than $700 ahead of Jews (Table 3.3B).

The argument that Catholicism impedes financial success in the United States should be considered closed; it does not. Nor does it interfere with educational attainment. In terms of mobility, Catholics are the second most mobile group in the society educationally, and the most

TABLE 3.3B

*Differences from Family Income Average of American Denominational
Groups in Metropolitan Regions Outside the South (1974
Dollars)—Non-Spanish-Speaking Whites Only*

DENOMINATIONAL GROUP	GROSS DIFFERENCES	NET DIFFERENCES
Jews	$ 2,295	$ 1,059
Catholics	1,211	1,782
Methodists	549	121
Presbyterians	277	−1,026
Episcopalians	− 145	−1,927
Lutherans	− 655	− 287
Baptists	−1,110	− 64

Average in metropolitan regions outside the South = $10,623

mobile denomination in annual family income. Only in occupational
prestige does there seem to be a Catholic lag. Catholics are not strongly
represented at the top occupational levels; indeed, they are not represented there as strongly as their educational attainment would seem to
entitle them. An explanation which seeks for some deficiency in the
Catholic belief system of community structure must contend with the
fact that deficiency has certainly not inhibited their considerable financial success, a success which cannot be explained by their disproportionate geographical location in cities in the north.*

The Catholic scores in income, education, and occupation reported
thus far are averages obtained when the earlier immigrants (Irish and
German Catholics) are combined with the more recent immigrants (southern and eastern European Catholics). Does a different picture emerge
when these averages are decomposed to the scores of the component
ethnic groups?

If one looks at Table 3.4A, one discovers that Irish Catholics have
the highest level of educational attainment of any Gentile white ethnic
group—though they are only one-tenth of a year higher than the British
Protestants. German Catholics are virtually at the national average of
11.5, and the eastern and southern European Catholics trail off to various
fractions of a year beneath the national average. However, turning from
educational stratification to educational mobility, we see that only three
of the ethnic groups report substantial educational mobility—the Polish
Catholics (1.2 years), the Italian Catholics (.8 of a year), and the Irish
Catholics (.6 of a year). The Irish are high on both educational stratification and mobility, while the southern and eastern Europeans are high

---

* Nor is there any change in the ranking or achievement in income, education,
and occupation if one focuses only on cities over two million. Unfortunately, the
number of respondents from those cities is too small to permit the mobility analysis
undertaken in this chapter.

TABLE 3.4A

*Differences from National Average in Educational Achievement
of Gentile White Ethnic Groups*

ETHNIC GROUP	GROSS DIFFERENCES	NET DIFFERENCES
Irish Catholics	1.0	0.6
British Protestants	0.9	0.1
German Catholics	0.1	0.2
Scandinavian Protestants	−0.1	0.0
German Protestants	−0.2	−0.2
Polish Catholics	−0.3	1.2
Italian Catholics	−0.4	0.8
Irish Protestants	−0.6	−0.3
"American" Protestants	−0.6	0.0
Slavic Catholics	−0.7	0.0
French Catholics	−0.7	−0.4
National average = 11.5		

SOURCE, tables 3.4A and B: Greeley, *Ethnicity, Denomination, and Inequality,* pp. 47, 48.

on mobility if not on stratification. If one moves from the national picture to the northern cities, very little is changed (Table 3.4B). British Protestants edge .4 of a year above the Irish Catholics in average education in such cities (12.8 for British Protestants, 12.4 for Irish Catholics). However, the Poles, the Irish, and the Italians continue to be the most

TABLE 3.4B

*Differences from Average in Educational Achievement of Gentile White
Ethnic Groups in Metropolitan Regions Outside the South*

ETHNIC GROUP	GROSS DIFFERENCES	NET DIFFERENCES
British Protestants(736)	0.8	0.0
Irish Catholics(653)	0.4	0.4
German Protestants(577)	0.0	−0.1
German Catholics(754)	−0.3	0.0
Irish Protestants(236)	−0.4	0.1
"American" Protestants(881)	−0.4	−0.3
Polish Catholics(410)	−0.5	1.2
Scandinavian Protestants(236)	−0.7	−0.8
Italian Catholics(820)	−0.9	0.5
Slavic Catholics(352)	−0.9	−0.1
French Catholics(257)	−0.9	−0.5
National average = 12.0		

mobile educationally. One can conclude that the Irish are responsible for raising the average in actual educational attainment, whereas the Poles and the Italians, and to some extent the Irish, are responsible for pulling up the relative attainment given parental education. There is no evidence of any barrier between the Catholicism of these groups and their educational achievement. Whether Catholicism stands in the way of graduate school training and academic careers is an issue which must await attention in Chapter 4.

Occupationally, three ethnic groups score above the national average (Table 3.5A)—British Protestants (43.9), Irish Catholics (43.7), and German Catholics (40.9) and Protestant (40.4). Three Catholic groups, Italians, Poles, and French, have scores above 39, and two Protestant groups, "American" and Scandinavians, are between 38 and 39. Finally, Slavic Catholics and Irish Protestants are at the bottom with scores of 37.6 and 36.7 respectively. If one moves into the cities of the North, Scandinavians move up substantially, but the Irish Protestants do not (Table 3.5B).

In terms of occupational mobility, the Polish Catholics, Irish Catholics and German Protestants have the highest achievement, although the Polish mobility seems to be almost entirely the result of geographic advantage, because it becomes average in the metropolitan regions outside the South. In these regions, incidentally, British Protestants continue to have both the highest occupational prestige and the greatest occupational mobility. Achievements of the Catholic groups in both educational stratification and mobility still leave them at a disadvantage in occupational competition with British Protestants of the same educa-

TABLE 3.5A

*Differences from National Average in Occupational Prestige
of American Gentile White Ethnic Groups*

ETHNIC GROUP	GROSS DIFFERENCES	NET DIFFERENCES
British Protestants	3.6	0.2
Irish Catholics	3.4	0.8
German Catholics	0.6	0.7
German Protestants	−0.3	−0.1
Italian Catholics	−0.8	0.4
Polish Catholics	−0.3	1.2
French Catholics	−1.0	0.3
"American" Protestants	−1.7	−1.5
Scandinavian Protestants	−2.0	−1.6
Slavic Catholics	−2.7	−1.2
Irish Protestants	−3.6	−2.6
National average = 40.3		

SOURCE, tables 3.5A and B: Greeley, *Ethnicity, Denomination, and Inequality*, pp. 51, 52.

TABLE 3.5B

*Differences from Average in Occupational Prestige of American
Gentile White Ethnic Groups in Metropolitan Regions Outside the South*

ETHNIC GROUP	GROSS DIFFERENCES	NET DIFFERENCES
British Protestants	3.1	2.2
Irish Catholics	2.4	1.0
German Catholics	1.5	2.6
German Protestants	0.9	0.8
Scandinavian Protestants	−0.6	0.2
"American" Protestants	−1.9	−1.2
Polish Catholics	−2.2	0.0
French Catholics	−3.0	−2.5
Italian Catholics	−3.1	−2.5
Irish Protestants	−3.3	−3.0
Slavic Catholics	−3.9	−0.5
National average = 41.8		

tional background in northern cities—though they have substantial advantages in the occupational competition over Irish and "American" Protestants.

Of particular interest is the case of Italian Catholics, who are second in educational mobility in the metropolitan North but second to the last in occupational mobility. Italians, in other words, do quite well educationally when compared with other Americans with similar educational backgrounds, but do quite poorly in competition for occupational prestige with others who have had the same educational experience. It is worth asking why this would be so. Perhaps Italian Catholics are simply not occupationally competitive, perhaps they are not ambitious for worldly success, or perhaps something else is at work.

There is no evidence that Italians are unambitious economically. The Italians are almost $2,000 above the national average in family income (Table 3.6A) and $454 above the northern metropolitan average (Table 3.6B), occupying second place behind the Irish Catholics nationally and third place behind the Irish and German Catholics in northern cities. The Irish are the most financially successful of all the white Gentile ethnic groups in the United States, and all the Catholic groups except the French do better than all the Protestant groups nationally; in the big cities in the North even the French earn about $200 more a year than do British Protestants. It was stated earlier in this chapter that whatever can be said about their occupation, Catholics were second only to Jews in income. We now observe that both in stratification and mobility, all of the Catholic ethnic groups seem to be more successful than all of the Protestant groups. Indeed, if we compare Table 3.6A with Table 3.3A we can observe that Irish, Italian, Polish, and German Catho-

TABLE 3.6A

Differences from National Family Income Average
for Gentile White Ethnic Groups (1974 Dollars)

ETHNIC GROUP	GROSS DIFFERENCES	NET DIFFERENCES
Irish Catholics	$ 2,473	$ 1,438
Italian Catholics	1,795	3,701
German Catholics	1,679	1,921
Polish Catholics	1,345	2,104
Slavic Catholics	873	2,802
British Protestants	401	−494
French Catholics	235	991
German Protestants	−195	−110
Scandinavian Protestants	−356	86
"American" Protestants	−679	271
Irish Protestants	−806	−129
National average = $9,953		

SOURCE, tables 3.6A and B: Greeley, *Ethnicity, Denomination, and Inequality,* pp. 54, 55.

lics all earn more in annual income than do Episcopalians. Furthermore, in metropolitan areas outside the South, Irish, German, Italian, Slavic, and Polish Catholics all have higher annual incomes than do any Protestant denomination (comparing Table 3.3B with Table 3.6B). Indeed, the Irish annual income is only a little more than $100 less than the Jewish annual income in cities outside the South. The myth of the blue-collar ethnic should by now be deader than a doornail.

But the question remains: Why do the Catholics show relatively lit-

TABLE 3.6B

Differences from Family Income Average of Gentile White Ethnic
Groups (1974 Dollars) in Metropolitan Regions Outside the South

ETHNIC GROUP	GROSS DIFFERENCES	NET DIFFERENCES
Irish Catholics	$ 2,170	$ 1,654
German Catholics	1,999	2,716
Italian Catholics	1,254	3,684
Slavic Catholics	900	2,476
Polish Catholics	813	1,427
French Catholics	465	1,712
British Protestants	259	−538
"American" Protestants	−233	17
Irish Protestants	−347	873
German Protestants	−425	−413
Scandinavian Protestants	−913	−756
National average = $10,623		

tle advantage in occupational mobility compared to their considerable advantage in educational and economic mobility? Irish Catholics have the best education and the best income in the country, and their educational and income mobility advantage over Protestants is substantial; still, in the metropolitan areas of the North, British Protestants have a higher rate of occupational mobility than do Irish Catholics (more than three and a half points higher in their standard deviation in occupational prestige). In other words, British Protestants get higher-prestige jobs than do Irish Catholics with the same education.

For the Italian Catholics the situation is even more paradoxical. Their educational and financial mobility is the highest in the country, but occupationally their mobility is less than average. In other words, Italians get more education than do other Americans with the same parental educational level and make more money on their jobs than do other Americans with the same educational background, but they do not get jobs that are nearly as prestigious. And the same is true for the Polish and Slavic Catholics. They make more money on their jobs than do people with comparable educational backgrounds, but they do not get nearly as good jobs. Again the question must be raised as to whether there is discrimination against these three groups at the upper levels of the occupational strata. It seems to be subtle and perhaps not all that harmful among the Irish, but rather blatant against southern and eastern European Catholics.

We have already noted that while the Catholic ethnics are more mobile educationally than anyone else in America, they are less successful than British Protestants or Jews in turning their education into occupational prestige. However, this lower prestige success rate does not seem to interfere with their income achievement, which is substantially higher than that of British Protestants. Interestingly (Table 3.7), the differences in occupational prestige are strongest among the Catholic

TABLE 3.7

*Differences from British Protestants in Occupational
Prestige of Jews and Irish, Italian, and Polish
Catholics by Education (with Blacks for Comparison)
(0-99)*

ETHNIC GROUP	DID NOT ATTEND COLLEGE	ATTENDED COLLEGE
Irish Catholics	0.63	− 1.35
Polish Catholics	− 1.43	− 4.09
Italian Catholics	− 1.80	− 3.93
Jews	2.63	2.95
Blacks	−12.90	−11.04

SOURCE: Greeley, *Ethnicity, Denomination, and Inequality,* p. 57.

ethnics who have attended college. Whatever factor that impedes Catholic occupational prestige, either internal weakness of the Catholic heritage or value system or external opposition, seems to be especially powerful among those Catholics who have had a college education. Thus Irish Catholics are .63 ahead of British Protestants in occupational prestige if they did not go to college, but fall 1.35 behind them (on a scale that runs from 0 to 99.0) if they attended college. For the Polish and Italian Catholics the difference from the British Protestant occupational prestige level more than doubles after college attendance. The Jewish lead over British Protestants is not affected by college attendance.

Access to higher occupational levels seems to be blocked especially for those Catholics who have had a higher education. This difference is not slight. For both Poles and Italians, the disadvantage is almost 40 percent of the disadvantage encountered by black college graduates (−11.04 for blacks, −3.93 for Italians, and −4.09 for Poles). Neither Catholic group suffers as much disadvantage as the black group, but the disadvantage is of sufficient magnitude to raise questions which must be asked in further research. If the black disadvantage is the result of discrimination—and few would deny it—then one must at least wonder if the Italian and Polish disadvantage is also the result of discrimination.

But however much they may be excluded (or exclude themselves) from the upper reaches of the occupational hierarchy, college-educated ethnics are hardly disadvantaged in income. Irish Catholic college graduates make $2,185 a year more than do British Protestant college graduates, Polish Catholics make $216 a year more, and Italian Catholics make $2,474 a year more, which makes them the second richest college graduate group in the country (after the Jewish $3,198 advantage over British Protestants).

There is, of course, the possibility that Catholics may settle for lower-prestige but higher-income occupations precisely because income is more important than prestige to those who still have the immigrant memory. To test such a hypothesis one would need to have available a much more detailed occupational coding than our composite sample provides.

At the very top levels of the prestige scale (categories running from 88 to 99 on the hundred-point scale) are 16 percent of the Jews, 9 percent of the Irish Catholics, 5 percent of the British Protestants, 4 percent of the Italians, and 2 percent of the Poles. But even in these top-prestige jobs, Jews earn more than Catholics (though only some $300 a year more than Irish Catholics), and Catholics earn more than British Protestants. Finally, non-Spanish-speaking Catholics are slightly overrepresented in the group making more than $25,000 a year. They constitute 20.6 percent of all Americans but 22.8 percent of those earning above $25,000.

It is on these findings, even more than the impressions I cited in the Preface, that I rest my case on the subject of a cultural division of labor in American society. On the basis of the evidence currently available, there is at least a reason to raise the question of whether Catholics, particularly those whose names end in vowels, are encountering barriers to the best jobs as a fulfillment of their educational and economic achievements. Clearly Catholics are not present in the upper levels of business, professional, and academic life. It is incumbent upon those upper levels of society to determine why Catholics are not among them, or they will run grave risk of being suspect of tolerating a cultural division of labor. Such toleration may have been justified in the past, when one could argue that Catholics were not ambitious enough or that their religion inhibited them intellectually, but the data presented here and subsequently in this book shatter both of these rationalizations. The possibility of systematic discrimination must be considered seriously.

Still, whatever the occupational discrimination, the Catholic immigrant groups have been immensely successful in America. The success of the Irish Catholics is surprising enough; one would not have thought of them as the richest and best-educated white Gentile group in America, but one might well have thought of them as comfortably and respectably middle class. The idea of the Italians being the second-richest Gentile group, and Poles and Slavs making more money in northern cities than British Protestants, will shock many. Poles, Italians, and other eastern European immigrants are late nineteenth- early twentieth-century arrivals who were not supposed to be able to acculturate easily to American ways. They are still thought of in the popular myth as the blue-collar ethnics, yet they turn out to be highly mobile educationally and extremely successful economically. Italians may still be thought of as sanitation workers, Poles as truck drivers or bartenders; it turns out, however, that they can laugh all the way to the bank—or perhaps all the way to the savings and loan. Whatever happened to the fatalism that was supposed to inhibit Italian achievement? Whatever happened to the dumb Pole of so many tasteless jokes? Whatever happened to the hard-drinking, hard-fighting Irish cop? Or, put more simply, *why* did it happen?

The Jewish immigrants were the products of two millennia of ancestors who had to live by their wits. They were craftsmen, tradesmen, inhabitants of small towns and cities. They were people of the book, believing firmly in education and practical learning. It is not difficult to see why they would be successful in the United States. But unlettered peasants who fled hunger, oppression, and injustice, and the rotting fields of western Ireland (bringing with them little besides their love of "the creetur," sexual puritanism, and their rigid, moralistic religion), the equally impoverished, amoral peasants from southern Italy, and the culturally disorganized Polish peasants—why have all these been so successful in America, while Baptists, Methodists, and even Episcopalians, who have been here for such a long time, have been less successful? It is an easy question to ask, but not easy to answer.

The success of the Catholic ethnics, particularly those from southern and eastern Europe, against all theoretical expectations, against the convictions of both sociologists and ordinary citizens, and indeed against their own self-image, is a baffling phenomenon. The Catholic immigrants came with little more than a dream of success and respectability. The ambition and industry generated by this dream seems to have been even more powerful than popular mythology and serious researchers thought possible. The Jews have made it big in the United States; that is not a surprise. So, too, have the Irish Catholics, and that is something of a surprise. That the Italians, the Poles, and the Slavs are in the process of making it big is an astonishment indeed.

But their success has not been recognized. Poles, Italians, Slavs, and even the Irish are not supposed to make it big. A dream doesn't come true in the full sense until others recognize and admit its truth. (As Mr. Dooley says, "Nothin' is iver officially true 'till a Repooblican admits it.") And then you are able to admit to yourself that it has indeed come true. This ultimate admission has not yet been conceded to the Catholic ethnics. If one wishes to find some explanation for the increased militancy among upper middle-class ethnics—particularly among those like Michael Novak, who are intellectuals—it is not so much economic and educational discrimination, perhaps not even so much the occupational discrimination that apparently still exists; it is rather a demand of admission from the rest of society that they have made it, that their dreams have come true despite the skepticism of society.

Alas for the Catholic ethnics. They have achieved worldly success precisely at the time when the cultural elite (having enjoyed its own success) has come to condemn educational, economic, and occupational success. It used to be said of the ethnics that they were no good because they could not achieve; now it will be said that they are no good because they have achieved. If you're ethnic, it seems, you lose even when you are winning.

Christopher Jencks and his colleagues, in their book on inequality,[9] argue that inequality not explained by the variables in their model can be attributed to luck. Daniel P. Moynihan, in a characteristic response, raised the question of why they did not attribute the differences to "pluck." The point of the witticism was serious. It may very well be luck whether you are born in a given religio-ethnic collectivity, but it certainly was more than luck that was required for the Jewish and Gentile turn-of-the-century immigrants to achieve not only parity but superiority in education, occupation, and income over the Protestant host culture in such a relatively short period of time. The immense energy and industry that went into this task can perhaps be attributed only to something as mystical as the power of a dream, a dream which has finally come true. Given the power of the dream and the abundance of its fulfillment, it is not difficult to see how the children, grandchildren, and great grand-children of the immigrants fail to share the contempt for American society which seems prevalent in many segments of the non-Catholic intelligentsia.

May it not be that for the descendants of the Jewish and Catholic immigrants there has been an "overthrust"? Could it be that the first age cohort of a population group that makes it in American society does so with such tremendous energy and such powerful "need for achieve-ment" that they not only do as well as everyone else, but better? (An economic version of Mr. Moynihan's now ancient comment that Harvard men would be investigated and Fordham men would do the investigating, perhaps?) May it be that in another generation or two the effect will wear off, and that the Catholic, Protestant, and Jewish ethnic groups will have relatively similar levels of achievement, while the blacks and the Spanish speaking begin to "thrust"? We simply do not have enough evidence to even begin to understand what has happened. And after a decade and a half of studying the phenomenon, I am sure that in my lifetime there will not become available resources to enable us to under-stand the truly remarkable performance of the groups that were sys-tematically excluded by the restrictive immigration legislation early in this century.

Peter Berger predicted "the blueing of America." He contended that if the sons and daughters of the elite went off to join communes and the counterculture, the management of American society would be taken over by the offspring of the "blue-collar ethnics," and the ethnics would become the richest and most successful people in America.

Berger was wrong to predict something like that for the future. It already has happened.

REFERENCES

1. Otis Dudley Duncan and David L. Featherman, "Psychological and Cultural Factors in the Process of Occupational Achievement," *Social Science Research* 1 (1972): 121–146; James S. Coleman et al., *Equality of Educational Opportunity* (Washington, D.C.: U.S. Office of Education, 1966); Peter M. Blau and Otis Dudley Duncan, *American Occupational Structure* (New York: Wiley, 1967).
2. See M. Bressler and C. F. Westoff, "Catholic Education, Economic Values,

and Achievement," *American Journal of Sociology* 49 (November 1963): 224–233; Gerhard Lenski, *The Religious Factor* (Garden City, N.Y.: Doubleday, 1961); Donald J. Bogue, *Population in the United States* (New York: Free Press, 1959); Galen L. Gockel, "Income and Religious Affiliation: A Regression Analysis," *American Journal of Sociology* 74 (May 1969): 632–647; R. W. Mack, R. J. Murphy, and S. Yellin, "The Protestant Ethic, Level of Aspiration and Social Mobility," *American Sociological Review* 21 (June 1956): 295–300; and B. C. Rosen, "Race, Ethnicity, and the Achievement Syndrome," *American Sociological Review* 24 (February 1959): 47–60.

3. Howard Schuman, "Free Will and Determinism in Public Beliefs About Race," *Trans-Action* (December 1969): 44–48.

4. A. J. Mayer and H. Sharp, "Religious Preference and Worldly Success," *American Sociological Review* 27 (April 1962): 218–227.

5. D. C. McClelland, *The Achieving Society* (New York: Van Nostrand, 1961).

6. J. Veroff, S. Feld, and G. Gurin, "Achievement Motivation and Religious Background," *American Sociological Review* 27 (April 1962): 205–217.

7. Lenski, *Religious Factor*, p. 101.

8. Ibid., p. 225.

9. Christopher Jencks et al., *Inequality: A Reassessment of the Effect of Family and Schooling in America* (New York: Basic Books, 1972).

CHAPTER

4

Catholics and
the Intellectual Life

THERE IS a simple, frequently advanced explanation for the absence of Catholics from the upper echelons of occupational prestige. Catholics are ambitious, so they go to school; Catholics work hard, so they make money; but unfortunately Catholics do not have the flair, the creativity, the fine sharp edge of intellectual curiosity and ingenuity that are required for eminence. Quite simply, Catholics may try harder, but they just aren't bright enough. There is something in their religion that seriously inhibits intellectual development.

Such an argument is not new. Kenneth Hardy has repeated it recently in an article in *Science*,[1] in which, ignoring the research evidence of the last fifteen years, he nonetheless contends that on the basis of a lower Ph.D. record from graduate schools in Catholic areas around the country, Catholicism is an intellectually inferior religion.*

The argument is apparently plausible. I have heard sociologists repeat it complacently after they have shaken their heads and observed that they don't know any Catholics among their colleagues—forgetting what the response of their hearers would be if they had substituted the words "blacks" or "women" for "Catholics" in their remarks. Indeed,

* Letters of protest which indicated Hardy's neglect of contrary evidence to be found in a number of articles in professional journals were systematically ignored by *Science's* editor.

the assumption of Catholic intellectual inferiority is strongly held and pervasive.

According to Lenski, the basic intellectual orientation of Catholicism as valuing obedience above autonomy, and also the Catholic tendency to value family and kin group above other relationships, creates a condition where "at both the conscious and subconscious levels of thought and action, membership in the Catholic group is more likely to inhibit the development of scientific careers than is membership in either Protestant or Jewish groups."[2]

Insofar as there is any empirical evidence for Catholic intellectual inferiority, it is based to a considerable extent on research done in the 1950s by R. H. Knapp and H. B. Goodrich. Two volumes resulted which are devoted essentially to the compilation of indices of the productivity of scientists and scholars by various colleges and universities.[3] In both studies the index of productivity from Catholic colleges was quite low. Catholic liberal arts colleges achieve 2.8 "scientists" per thousand graduates, while Catholic universities had an index of 1.7.[4] In the study of younger scholars sixteen Catholic men's colleges had an index of 1.96, as opposed to 10.85 for twenty-three non-Catholic men's colleges.[5]

Two quotations from the Knapp and Goodrich volumes summarizes their explanation of this phenomenon:

The conspicuously inferior position of virtually all Catholic institutions in the production of scientists apparently requires . . . explanation. . . . Without wishing to pose any final interpretation of the low production of Catholic institutions, we may at least present the following considerations which have been advanced in partial explanation. First, Catholic institutions are in good part concentrated in the eastern sections of the United States, a region not noted for the high production of scientists. Second, the Catholic population of America, from which these institutions have largely drawn their students, has come from parent European cultures not conspicuous in recent times for scientific accomplishments. Third, Catholicism has permitted comparatively little secularization of outlook among its constituents and has maintained a firm authoritarian structure. Fourth, Catholicism has been a consistent opponent of physical monism, that philosophic tradition under which science has for the most part advanced. However, though this class of institutions has made small contributions to the ranks of American scientists it is noteworthy and probably significant that its contributions to the legal profession . . . have been exceptional.[6]

We expected that Catholic institutions would be marked by relatively large contributions to the field of humanities. In this speculation, however, we were again mistaken. Catholic institutions, although exceptionally unproductive in all areas of scholarship, achieve their best record in the sciences.[7]

Perhaps the most interesting body of comment on the antiintellectualism of American Catholics is the theory or demitheory which was evolved in the 1950s by Catholic self-critics and has yet to be abandoned by them. Thomas O'Dea sums up the position of these self-critics: "The fact is that although American Catholics have the largest and the most expensive educational system of any Catholic group in the world, a genuine Catholic intellectual life is still dependent upon the translation of European works."[8] And Gustave Weigel puts the matter even more succinctly: "The general Catholic community in America does not know what scholarship is."[9]

There seem to be at least eight elements in the theory which the self-critics have evolved to explain the intellectual weakness of American Catholicism. The first point made is that the generally authoritarian approach of American Catholicism inhibits the development of scientific curiosity. In Karl Herzfeld's words, "There is no conflict between the *results* of science and the *doctrines* of the Church, but there is in my opinion a conflict due to training method."[10]

A second aspect of the explanatory theory is the clerical domination of American Catholic life. Weigel says:

Another common persuasion among our people is that smart boys should go into the priesthood. Those students who distinguish themselves in high school are always asked if they intend to study for the priesthood. Actually the seminaries get a good percentage of them. What is not recognized, often not even by priests, is that priestly formation need not be scholarly nor is scholarship its true aim.[11]

And O'Dea adds even more forcefully:

The intellectual life is not likely to grow up in a situation where community is transformed into bureaucracy and where a clerical monopoly tends to damage the quality of clerical leadership itself. If the laity remain the feet and the clergy the head in terms of constructive thinking, then we shall perpetuate a situation that inhibits the development of intellectual responsibility and creativity on the part of all. Such a condition not only inhibits the development of intellectuals but drives potential intellectuals away from and often into antagonism to the Church.[12]

A third explanation would suggest that the concern of the church with the next world leads it to place little value on the present life and to have little concern for the temporal order which is the object of scientific research. It is assumed that the piety of the largely Irish-dominated American Catholicism is not especially interested in the transitory events of this vale of tears.

A fourth explanation suspects that there is actually a fear of scientific knowledge within the Catholic community. In O'Dea's words:

The American Catholic community not only fails to produce an intellectual elite, but even tends to suspect intellectuals because American Catholics sense the relation of the intellectual to these two partially repressed American problems. The intellectual symbolizes in some way the facing of crises and the challenges of uncertainty, as well as a critical attitude toward the accepted values of middle-class life, especially those which are most crassly materialistic. Catholics, who pride themselves on their spirituality, are more prone to support anti-intellectualism than are many other of their countrymen. They see the intellectuals as those who seek to reveal the basic existential ambiguities that they themselves wish to evade. They often project this dilemma onto current politics and see such intellectuals—who are in our culture . . . often liberals— as "disloyal." One sometimes feels that for some Catholics "loyalty" is so supreme a virtue that it would be disloyal to ask, "loyalty to what?"[13]

O'Dea rejects the possibility that the explanation for the lower levels of Catholic intellectuality may be found in the condition of Catholics as recent immigrants—an explanation that should have occurred to sociologists, and should not have been rejected without testing. The formalism, authoritarianism, clericalism, moralism, and defensiveness of Catholics, according to O'Dea, are the basic milieu which inhibit the development of mature intellectual activity.

Daniel and Sidney Callahan comment:

Surely the Catholic college which says, in effect: you are intellectually mediocre, you need our strict control to mature and to keep your faith, and it might be harmful if you were left to your own devices in your newspapers and student groups, cannot be said to foster a spirit of engagement and responsibility. . . . One might well ask if the general run of Catholic education when not inspired by a desire to develop personal insight and integrity, does not in fact simply indoctrinate in the bad sense of the word.[14]

And Joseph Cuneen concludes that "both training in the techniques of learning as well as disinterested top level scholarship suffer from understandable but misguided desire to have the educational content itself provide the direct motives of piety."[15]

Perhaps the most obvious element in the explanation is the immigrant status of the American Catholic group. Some of the bolder of the self-critics have felt that this condition was probably at the root of most of the other problems, and that when and if American Catholics left their immigrant mentality behind, there would be the beginnings of an intellectual development.

Monsignor John Tracy Ellis, in a careful historical study, suggests that part of the explanation must be the absence of a tradition of scholarship among American Catholics, due at least in part to the socioeconomic difficulties of the immigrant groups.[16] Ellis' work is probably the most influential of all the comments made on American Catholic intellectualism.

Closely related to the immigrant origins of American Catholicism is the presumed materialism of the descendants of the immigrants who are attempting to "succeed" in American life. Says Herzfeld, "It seems, therefore, that a much higher percentage of Catholic than of non-Catholic students stop their studies after college to go into business."[17] In this view Catholics are too busy trying to make money to be concerned about being scholars.

Finally, it is often suggested that one of the causes of American Catholic antiintellectualism is the ghetto attitude of defensiveness which is characteristic of American Catholics. Weigel argues:

American Catholicism until very recently has always had the feeling of being a beleaguered community. An ubiquitous, formidable enemy was threatening its very existence. Loyal defense was needed, not a divisive effort of criticism. Everything that was, took on a whole aspect; to be loved and died for. Such an atmosphere was not propitious to American Catholic intellectualism.[18]

The self-critics have not had much concrete data to back up their positions. Ellis cites a considerable number of studies, although most of them date to the pre-World War II era. Herzfeld mentions the Catholic percentage at Johns Hopkins and the fact that there are no American Catholics who have won Nobel prizes in physics. McKernan, McCluskey, and the editors of *America* discuss the paucity of National Science Foundation, Rhodes, and Fulbright scholarships for American Catholics.[19] References are also made on occasion to the Knapp and Goodrich data (which would again apply to a generation coming out of college before World War II).

There were hints available before 1960 that those both inside and outside of Catholicism who were convinced of the intellectual inferiority of Catholics were talking about a situation which existed perhaps before World War II, and perhaps even into the early 1950s, but which no longer applied. In a study done in the late 1950s, James Davis and his colleagues found that one-quarter of the arts and sciences graduate students in the country were Catholics—exactly the same proportion as Catholics in the general population.[20] Similarly, Seymour Warkov and I discovered that in the middle 1950s graduates of Catholic high schools had suddenly become more likely than graduates of other high schools to go on for college degrees.[21] In retrospect, it is possible to see both these shifts as part of the post-World War II change in the southern and eastern European ethnic groups that we have recorded in Chapter 2. At that time, however, they were straws in the wind which were not taken seriously. In 1961 a very large representative national sample survey of college graduates conducted by NORC provided an opportunity to "for the love of heaven find out why Catholics aren't going to graduate schools" (as then superintendent of Catholic schools in Chicago and now Auxiliary Bishop William E. McManus put it). Alas, the question could not be answered, because they *were* going to graduate school.

Seven hypotheses derived from the various works on Catholic antiintellectualism mentioned above were tested:[22]

1. Catholics will be less likely to go to college.
2. Catholic graduates will be less likely to go to graduate school.

3. Catholics who go to graduate school will be less likely to choose the arts and sciences, that is, the academic fields.
4. Catholics in the academic fields will be less likely to go into the most scientific of sciences, that is, the physical sciences.
5. Catholics who go into the academic fields will be less likely to plan a research career.
6. Catholics who go into the academic fields will be less likely to be religious and more likely to be apostates.
7. Catholics will tend to overchoose large corporations as employers, business as an occupation, and security and the avoidance of high pressure as occupational values.

All the hypotheses collapsed in the face of the data. Catholics were 18 percent of the college graduates in the country but 25 percent of the whites who graduated in June 1961—precisely their proportion in the national population. Forty-seven percent of the Jews, 33 percent of the Catholics, and 28 percent of the Protestants went on to attend graduate school in the following year. Thus, the second hypothesis collapsed. Forty-six percent of the Catholics who were going to graduate schools, as opposed to 43 percent of the Protestant, planned graduate studies in the arts and sciences; the third hypothesis collapsed. Thirty-two percent of the Catholics, 33 percent of the Jews, and 30 percent of the Protestants planned careers in the physical sciences; the fourth hypothesis collapsed. Three-fifths of both the Catholics and the Protestants who were going to be scholars planned careers which would involve research, and the fifth hypothesis collapsed. The Catholic academics were not less likely to go to church, and the sixth hypothesis collapsed. Catholics did indeed overchoose large corporations, but they did not overchoose business as an occupation or security and avoidance of high pressure as occupational values, and thus two-thirds of the final hypothesis collapsed.

The data in the 1961 NORC study did not necessarily refute the arguments of Catholic antiintellectualism, but they provided a massive "nonconformation" of these arguments that should have given those who proposed them at least some pause.

A storm of controversy swirled around the 1961 research. D'onald Warwick wrote a letter to the *American Journal of Sociology* protesting that all that was being analyzed was the 1961 graduate school plans of college graduates at the time of graduation.[23] Lenski was even more forthright:

Perhaps the key to the explanation of Father Greeley's unique and intriguing findings is to be found in the fact that they deal with the intentions and aspirations of young people on the verge of graduate school whereas the other studies . . . deal with established men of science. The findings of our study suggest that in general the aspirations of Catholics are as high as those of non-Catholics. . . . Catholics are ambitious and hard working. If they fail to get ahead in certain areas, it seems to be due more to the influence of other factors whose importance is easily overlooked. . . .[24]

The outcry from Catholics was even more outraged. John Donaven accused the NORC report of being "naive empiricism," arguing that the "free-wheeling, critical, creative, and speculative bent of mind that marks the intellectual" is probably lacking in Catholic respondents to the NORC study. They were rather merely "intelligent graduates of the college population."[25] And James W. Trent suggested that these young people would be "authoritarian, and intellectually docile graduate students who would contribute little more to the flow of intellectuality and creativity than the ordinary high school graduate."[26] Trent also suggested that the NORC sample frame was deficient. In fact, he circulated a memo arguing this privately to a considerable number of Catholics, without bothering to find out whether he understood accurately the NORC sample design—in its own way an antiintellectual exercise which seemed to make his own case better than he knew. Daniel Callahan was unimpressed by the NORC findings: "The real culprit is the American Catholic mentality."[27] A number of Catholic editorial writers rushed to the defense of John Tracy Ellis and his classic critique of Catholic intellectualism in the United States, "American Catholics and the Intellectual Life." They overlooked the fact that Ellis' work was historical, and the data on which he was reporting described a situation which had existed thirty years previously.

The fury of the response, particularly from among the Catholics themselves, was intriguing. If one could not prove that the Catholic respondents in the NORC study were "free-wheeling, . . . creative, and speculative" and capable of a contribution beyond that of an ordinary high school graduate, neither Trent nor Callahan were capable of proving the opposite. The NORC report had made clear that "intellectualism" was being used in a restrictive sense.

The use of the word "intellectualism" in this paper needs to be clarified. It is clear that going to graduate school in the physical sciences and planning a career of research in this area, for example, are not necessarily indicators of intellectuality, much less of potential scholarship. It is argued merely that the first six hypotheses proposed above would seem to follow logically from those writings that question the intellectuality or at least the orientation toward science of American Catholics. If the hypotheses are not supported by the data, previous researches are not necessarily disproved, but at least must be seriously re-examined.[28]

But the critics were not even ready to concede that. On a priori grounds they knew that the Catholics in the 1961 study would prove not to be "free-wheeling, critical, creative, speculative" and that they would be "authoritarian and intellectually docile." Lenski knew further that they would not continue their graduate studies.

Fortunately for the advancement of human knowledge, if not for the position of the various critics, the NORC survey was a panel study, and by 1967 information was available on how the 1961 graduates were doing in their career plans.[29] Table 4.1 would suggest that none of the major conclusions reported on the 1961 study needed to be modified.

TABLE 4.1

*Graduate School Experiences by Religion of June 1961 College Graduates
(Only White Males from Upper Half SES Backgrounds Who Grew Up in
New England or Middle Atlantic Cities with a Population of Over 500,000)* *

GRADUATE SCHOOL EXPERIENCES	PROTESTANTS	CATHOLICS FROM CATHOLIC COLLEGES	CATHOLICS FROM OTHER COLLEGES	JEWS
Percentage still in graduate school (spring 1964)	45	46	44	60
Percentage with M.A.	12	15	11	24
Percentage expecting Ph.D.	21	20	15	26
Percentage expecting academic careers	20	19	15	43
Percentage in arts and sciences graduate programs	20	22	18	30
Percentage of those in graduate school who attend full time	58	57	38	55
Of those expecting Ph.D., when it is expected (percentage):				
By 1965	38	28	26	26
By 1967	78	79	62	89
Percentage of Ph.D. topics chosen	65	70	55	68
N	163	510	316	121

SOURCE: Andrew M. Greeley, "Religion and Academic Career Plans: A Note on Progress," *American Journal of Sociology* 72 (May 1967): 669. (This and following table are reprinted by permission of the publisher, The University of Chicago Press — © 1967.)

*Subsample includes all respondents whose original religion was Catholic and one of every six whose original religion was not Catholic.

It should be noted that the percentages in Table 4.1 were based only on white males from the upper socioeconomic half of the population in the New England and northeastern section of the country who were raised in cities with a population over a half million. Thus, there was built into the table control for race, sex, socioeconomic status (SES), region of the country, and size of city of origin.

While Jews were still considerably different from Gentiles, differences between Protestants and Catholics who attended Catholic colleges were almost nonexistent. Better than two-fifths were still in graduate school, close to three-fifths were attending school full time, between one-sixth and one-seventh already had their M.A., and about one-fifth were expecting to obtain a Ph.D. Almost four-fifths felt that their doctoral work would be finished by 1967, and better than two-thirds had already chosen a topic for their doctoral dissertation, while two-fifths of those had chosen their doctoral topics reported they had done so independently and not at the urging of a particular advisor. Finally, one-fifth of both of those groups expected academic careers and were specializing in arts and sciences disciplines while they were in graduate schools.

The deviant groups, insofar as there was any deviancy in Table 4.1, were the Jews and, to some extent, the Catholics who attended non-Catholic colleges. The former were much more likely to be in graduate school, to have obtained their M.S., to have expected their doctorate before 1967, to have expected academic careers, and to have specialized in the arts and sciences. The latter were considerably less likely to have gone to school full time and were somewhat less likely to have expected the doctorate or to have planned academic careers; among those in this group who were pursuing the Ph.D., there was less of an expectation of finishing the doctorate by 1967.

Of those respondents among the three Gentile religious groups which were analyzed who were going to graduate schools in the arts and sciences, approximately one-fifth of each group was attending one of the top twelve graduate schools. Table 4.2 provides information on how the graduate students in the quality graduate schools fared in the pursuit of their aspirations. While the case base in Table 4.2 was admittedly relatively small, at least the proportions in the table provide grounds for interesting speculation. None of the four analytic groups was likely to have dropped out of school since the spring of 1962. The Catholic students were slightly more likely to report that they had an A grade-point average. Only the Catholics who went to non-Catholic undergraduate colleges were different from the others in reporting plans for the doctorate, and the Catholic school Catholics did not lag behind the Jewish or Protestant groups in their plans to obtain their Ph.D. by 1965 or 1966. Protestants and Catholic school Catholics were much more likely to have had their topics approved.

Even though the small number of respondents represented in Table 4.2 make it impossible to use more detailed socioeconomic controls, there is surely no evidence to be found in the table for the notion that Catholics (at least if they had gone to Catholic colleges) were unsuccessful in the high-quality graduate schools. Half of the Protestant graduate students and half of the Catholic graduate students who did not attend Catholic

TABLE 4.2

*Graduate School Experiences by Religion of Arts and Sciences Students
from the June 1961 Class Who in the Spring of 1962 Were Attending
Top Twelve Graduate Schools (Whites Only)* *

GRADUATE SCHOOL EXPERIENCES	PROTESTANTS	CATHOLICS FROM CATHOLIC COLLEGES	CATHOLICS FROM OTHER COLLEGES	JEWS
Percentage still in graduate school (spring 1964)	95	100	100	88
Percentage with A grade-point average	10	16	17	14
Percentage planning Ph.D.	97	98	66	100
Of those expecting Ph.D., when it is expected (percentage):				
By 1965	59	56	33	24
By 1966	79	96	47	86
Percentage having thesis topic *approved*	50	59	22	19
Percentage still in religion in which they were raised	55	85	52	71
Percentage still in religion in which they belonged at college graduation	81	98	54	79
N	40	54	27	21

SOURCE: Greeley, "Religion and Academic Career Plans," p. 671.

*Subsample includes all respondents whose original religion was Catholic and one of every six whose original religion was not Catholic.

undergraduate colleges were no longer members of the religion in which they had been raised, while some 15 percent of the Catholic school Catholics had defected, and almost 30 percent of the Jews no longer considered themselves Jews. However, for all but the non-Catholic school Catholics these decisions seemed to have been made before the graduate training began. The vast majority of the other three analytic groups reported that their religious preference after three years of graduate school was no different from their religious preference at the time of graduation. Only the Catholics who did not go to Catholic colleges seemed to have delayed their decision to leave Catholicism until graduate school.

Thus there is no evidence from the 1967 data that American Catholics were disinclined to enter the top-quality arts and sciences graduate departments, nor that they did poorly in their academic efforts in these departments (especially if they attended Catholic undergraduate colleges). Finally, those who had Catholic undergraduate training did not seem to have found any conflict between religious faith and academic pursuits.*

Catholics, particularly those who had attended Catholic college, had not been inhibited from achieving some measure of moderate success in the early and middle years of their graduate school training. By the middle 1960s Catholic graduate students were working hard, getting good marks, making progress toward their degrees, and more than likely to have their theses approved. There was little reason to assume that these young people would lose their enthusiasm for the academic life.

But the NORC research continued, and in 1968 the fifth wave of interviews with the 1961 graduates was completed.[30] The questions asked in this wave enabled respondents to answer not just the question of the impact of Catholicism, but also of the impact of Catholic education. If Catholicism is a religion which tends to produce an antiintellectual orientation, then Catholic education should reinforce such orientation.

Catholic education did not seem to interfere with educational or financial success or with the choice of an academic career (Table 4.3). The Catholics who had all their education in Catholic schools were twice as likely as those who had all their education in public schools to have had either a terminal professional degree or a Ph.D. They were also the most likely to be working for a college or university or for a research organization, though the differences among the six groups were too small to be important.

Those who had all their education in Catholic school were more likely than any other respondents in the sample to score high on the reading index,† although Catholics with partial Catholic education scored lower on this measure than did non-Catholic respondents.

Finally, with the exception of those who went to Catholic colleges after public grammar school and high school, there was a positive relationship between Catholic education and earning more than $11,000 per year.

Thus there was no evidence that attending Catholic schools interferes with one's intellectual, educational, or financial achievement—quite the contrary. If anything, Catholic school attendance seemed positively to facilitate such achievements. It may be that those who attended Catholic schools came from families where there was more emphasis on upward mobility and educational achievement than there was in other families

---

* While the accusation of bad sampling in the first NORC study was ludicrous, a second study done by Michael Schiltz in 1964, using an entirely different sampling frame, merely duplicated the 1961 findings, thus putting to rest the charge of bad sampling methodology leveled against one of the country's oldest research institutions.

† Based on reported ownership of books and frequent reading of nonfiction and poetry.

TABLE 4.3

Catholic Education and Achievements in 1968

(percentage June 1961 college graduates)

EDUCATION AND ACHIEVEMENT	ALL CATHOLIC	CATHOLICS				NON-CATHOLICS
		CATHOLIC COLLEGE ONLY	CATHOLIC PRIMARY ONLY	CATHOLIC PRIMARY AND SECONDARY	ALL NON-CATHOLIC EDUCATION	
Ph.D. or professional degree	18.0	10.0	15.0	14.0	9.0	12.0
Ph.D.	5.8	0.0	2.1	1.9	3.7	3.8
Earning more than $11,000	37.0	27.0	40.0	45.0	32.0	27.0
High on reading scale	28.0	18.0	19.0	7.0	16.0	23.0
Choosing academic careers	29.0	20.0	17.0	13.0	19.0	27.0
N	459	132	235	101	449	6,289

SOURCE: Andrew M. Greeley, "Continuities in Research on the 'Religious Factor,'" *American Journal of Sociology* 75 (November 1969): 355. (This and following table are reprinted by permission of the publisher, The University of Chicago Press — © 1969.)

(a curious reversal of the Protestant ethic). It also may be that they were more successful because they enjoyed greater emotional well-being inside the subculture than they would outside the subculture. Finally, it may also be that faculty and administrators of Catholic colleges put strong emphasis on academic and economic achievement in the training of their students. There is some evidence of this.[31] However, it seems it is clear that it is definitely incorrect to call the rather surprising scores of Catholics on measures of "intellectualism" in NORC's June 1961 sample a function of great aspirations that would not be carried to fulfillment. The graduates of Catholic colleges cannot be written off as "antiintellectual."

But the most striking data to be reported in 1969 may be observed in Table 4.4. Attendance at Catholic college, particularly after attendance at Catholic grammar school and high school, correlated strongly with liberal, not to say radical, political positions. Those who attended Catholic colleges (independent of their primary and secondary education) were more likely to describe themselves as liberal Democrats, though the differences on this item are much less striking than they were on the other four items of the table. Those who attended Catholic colleges were considerably more sympathetic to black militants and also substantially more likely to accept the conclusion of the Kerner report that riots are caused by white racism. Even more surprisingly, those who attended Catholic colleges after Catholic grammar and high school experience were far more sympathetically disposed toward student militancy than were other Catholics who graduated from college, and they were slightly more favorably disposed toward militancy than the typical American college graduate.

The 1968 research fairly well demolished the Lenski prediction that the young Catholic respondents to our survey would not persist to the Ph.D. or choose academic careers. It also cast grave doubt on the Trent suggestion that they were not likely to engage in uniquely and specifically intellectual activities. They read more and were likely to be more critical of American society, two reasonably good indicators that one is exercising the authentic role of the intellectual.

Nothing more was heard from Lenski, Trent, O'Dea, Callahan, or Ellis.

Unfortunately, it has not been possible to continue to pursue the 1961 graduates. However, a reanalysis of the 1968 data that focused on members of the physical and biological sciences and looked at the different ethnic groups represented in those disciplines offered some interesting data (Table 4.5). The Irish Catholics were ahead of the other ethnics, including the Jews, on all the items. German and Polish scientists compare favorably with the three Protestant groups represented on the table in the proportion having a Ph.D., working in a university or

TABLE 4.4

Social Attitudes and Catholic Education in 1968
(percentage June 1961 college graduates)

| SOCIAL ATTITUDES | ALL CATHOLIC | CATHOLICS | | | | NON-CATHOLICS |
		CATHOLIC COLLEGE ONLY	CATHOLIC PRIMARY ONLY	CATHOLIC PRIMARY AND SECONDARY	ALL NON-CATHOLIC EDUCATION	
Percentage liberal Democrat	23	36	17	14	18	16
Percentage thinking Negro protests healthful for America	63	68	53	47	53	53
Percentage blaming riots on white racism	47	41	33	28	33	33
Percentage supporting student political involvement	31	16	21	24	23	28
Percentage supporting student militancy	29	20	17	13	19	27
N	459	132	235	101	449	6,289

SOURCE: Greeley, "Continuities in Research on the 'Religious Factor,'" p. 672.

TABLE 4.5

Attitudes and Behavior of Younger American Scientists by Religion and Ethnicity*

	Ph.D. (%)	CAREER IN UNIVERSITY OR RESEARCH (%)	EXPECTING MAJOR SATISFACTION FROM CAREER (%)	HIGH ON "READING FREQUENCY" INDEX (%)	DESCRIBING SELF AS INTELLECTUAL (%)	WEEKLY CHURCH (%)
Protestant						
Anglo-Saxon (295)	19	24	28	15	47	28
German (181)	15	21	31	10	48	33
Scandinavian (27)	22	29	28	10	52	44
Catholic						
Irish (40)	43	58	51	32	75	73
German (66)	29	12	40	12	65	58
Italian (31)	10	6	15	4	33	51
Polish (34)	23	37	6	2	44	91
Jewish (27)	27	30	25	25	64	10

SOURCE: Andrew M. Greeley, "The 'Religious Factor' and Academic Careers: Another Communication," *American Journal of Sociology* 78 (March 1973): 1253. (This and following tables from this source are reprinted by permission of the publisher, The University of Chicago Press — © 1973.)

*Those who in 1968 reported that their long-range career would be in the physical, chemical, or biological sciences or in mathematics.

research center, and describing themselves as intellectuals. The Irish had made it big in the world of the academy, and the other ethnics did not lag far behind.

A 1969 study sponsored by the Carnegie Commission on Higher Education[32] enabled us to pursue the 1961 graduates one step further. While the college faculty under thirty interviewed in 1969 may not have been the same respondents we interviewed in the NORC studies, they were nonetheless a sample of members of the same cohort that had pursued academic careers. Using the technique of cohort analysis demonstrated in Chapter 3, we can see whether the proportion of Catholics in university faculties, particularly in the thirty "elite" colleges that were focused on in the Carnegie report, had changed through the decades (Table 4.6).

TABLE 4.6

Religious Background (Religion in Which Raised) of American Professorate in 1969 by Age—Elite Colleges and Universities Only

	N	PROTESTANT (%)	CATHOLIC (%)	JEWISH (%)	OTHER AND NONE (%)	TOTAL* (%)
65 and over	439	75	9	9	7	100
60-64	1,012	73	9	10	7	99
55-59	1,317	70	10	14	6	101
50-54	1,756	66	11	16	7	101
45-49	2,425	58	12	20	10	100
40-44	2,749	55	12	23	11	101
35-39	3,322	54	15	20	11	100
30-34	3,398	54	15	20	10	99
29 and under	2,673	52	19	21	9	101

SOURCE: Greeley, "The 'Religious Factor' and Academic Careers," p. 1249.

*Totals are not always 100% because of rounding errors.

Catholics were only one-tenth of the faculty who were over fifty in such colleges and universities. Thus, in the generation of those who had attended college before World War II, only about one-tenth of those who became faculty members at the best colleges and universities in the country had been raised Catholic. However, when one turns to the generation who grew up during the Cold War, one-fifth of the faculty at these colleges and universities were Catholic. Catholics, then, were approximately equal numerically to the Jews and approximately equal proportionately to the Protestant population (with relation to their respective proportion in the total American population) on the elite faculties. The breakthrough for Jewish faculty members at the elite schools apparently took place for those students who were in graduate school between 1935 and 1950. Since that time, Jews have been consistently about one-fifth of the elite faculties. The breakthrough for Catholics (that is, for those faculty members who are presently under forty) seems to have begun in about 1955—at the time, James Davis reported that one-quarter of the arts and sciences graduate students in America were Catholic.[33]

Nor were the Catholic faculty under thirty-five, whether at elite or nonelite colleges (Table 4.7), inferior to the Protestant faculty in either acquisition of tenure or the publication of five or more scholarly articles—though both groups were substantially behind the Jews in both these respects.[34] As a matter of fact, the faculty with Catholic backgrounds differed from their Protestant colleagues only in frequency of church attendance, and are similar in their attitudes, self-definition, and values at both elite and nonelite colleges.

It has been an intricate and perhaps tedious affair to trace the path of the 1961 graduates through the decade after their graduation. But there has been no other way to document solidly the fact that a younger generation of Catholics have found no real obstacle between their religion and successful pursuit of academic careers. They have gone to graduate school, obtained their doctorates, have appointments in the elite colleges and universities, are publishing articles, displaying the values and attitudes appropriate to intellectuals, and were, at least before the bottom fell out of Catholic religious practice, not substantially less likely to go to church than any other Catholic of their generation. And all of these things were more likely to have occurred if the young men and women in question had gone to Catholic colleges; they were still more likely if all their education prior to graduate school was Catholic.

No argument is being made here that this was always the case. On the contrary, the evidence presented in Chapter 2 suggests that this only began to be the case in the years after World War II, when the descendants of the southern and eastern European ethnic groups began to be acculturated into American society. By the middle and late 1940s these groups were beginning to approach parity in college education, and by the middle 1950s they had begun to approach parity in graduate school attendance. There was a monumental change in the Catholic collectivity between 1945 and 1965, but the change had nothing to do with the Second Vatican Council, and was accomplished more by the crossing of the various acculturation barriers of the southern and eastern European immigrants—college attendance, graduate school attendance, intellectual careers. There is no evidence that these young people found their Catholic faith, Catholic family life, or Catholic education a handicap in breaking through the barriers. It would appear, actually, that Catholic education was a positive asset, and that in fact the more Catholic education they had, the more effectively they would crack the barriers. There may still be a subtle, as yet unisolated, element of Catholic antiintellectualism which the research of the last decade and a half overlooked;

TABLE 4.7

Percentage of Younger Professors in Colleges and Universities
by Religion Having Certain Attitudes and Behavior

	29 AND UNDER			30-34		
	PROTESTANT	CATHOLIC	JEW	PROTESTANT	CATHOLIC	JEW
	ELITE					
With tenure	2	3	3	23	22	31
5 or more professional articles published	15	14	17	40	39	51
Describing self as "intellectual" (strongly agree)	22	23	34	23	27	33
Supporting Vietnam withdrawal	28	30	43	24	22	31
Sympathetic with student protest	61	58	69	56	54	61
Attending cultural event 2 or 3 times a month	34	33	37	31	30	36
"Left" politically	12	13	21	11	10	14
Church attendance at least monthly	32	52	8	35	53	8
	NONELITE					
With tenure	3	2	2	20	19	21
5 or more professional articles published	1	1	1	5	5	12
Describing self as "intellectual" (strongly agree)	13	16	26	14	14	31
Supporting Vietnam withdrawal	20	20	39	14	17	35
Sympathetic with student protest	46	50	57	43	49	61
Attending cultural event 2 or 3 times a month	26	31	36	26	26	35
"Left" politically	6	8	20	5	5	16
Church attendance at least monthly	49	63	11	57	70	13

SOURCE: Greeley, "The 'Religious Factor' and Academic Careers," p. 1251.

but those who believe in such a "deep" flaw in the Catholic personality have the burden of defining that flaw, making it operational, and showing that it still exists even among those practicing Catholics who are publishing faculty members of the elite colleges and universities.

I do not know which of the thirty colleges and universities that were labeled "elite" contain Catholic faculty members; however, I have a strong impression that Catholics are not working in the great private universities of the country. They are much more likely to be found on the faculties of the better state universities which make it into the top thirty "elite." It may be that the young Catholic scholars are very good, but just not quite good enough to make it into the really big time. They lack the "flair," the "extra touch" that distinguishes between competence and brilliance and which characterizes the faculties of Harvard, Yale, or Chicago.

Such a possibility cannot be refuted on a priori grounds, but it still must be noted that we are dealing with a limited aspect of intellectualism—albeit, perhaps, the most important aspect. If it is conceded, as it must be in the face of data, that Catholics are very good intellectually, then one cannot explain their underrepresentation in the upper echelons of the occupational prestige world as described in Chapter 3. "Flair" is a relatively scarce commodity; there is simply not enough of it around to explain notable differences in prestige scores. Hence, there does not seem to be any explanation for the lower occupational prestige of Catholics, given their higher educational and income achievements, other than some subtle (and perhaps not so subtle) forms of discrimination. Catholic intellectual deficiencies, if they exist, are simply not large enough to explain the variation of prestige scores.

But how much "flair" is there really in the upper reaches of the elite academies? Have you read recently any of the scholarly publications of the social sciences? How much flair do you encounter there? Have you attended a faculty cocktail party or dinner party at Harvard or the University of Chicago? How many brilliant people, really preeminent scholars in their fields, do you encounter? How much imagination, creativity, and free-wheeling speculative intellect does there seem to be around the great universities? And is it not the case that even at those schools, junior faculty members with flair and ingenuity are likely to have serous trouble staying on? I would observe that if Catholics do not make it into the faculties of the best private universities of the country because they do not have flair, then a hell of a lot of other people don't belong there either.

In retrospect, one wonders why none of the discussion of the alleged intellectual inferiority of Catholics has raised two critical questions. Why has the entire discussion seemed to have taken for granted that Catholics and Protestants began the quest for intellectual excellence on exactly the same socioeconomic base? It was as though the turn-of-the-century immigration and the struggle of the southern and eastern European to achieve acculturation had never occurred.

Why has the question never been raised as to whether the absence of Catholics in the upper reaches of the academy could be to some extent the result of discrimination instead of Catholic inferiority? Why have the assumptions of black and female inferiority been questioned so vigorously, while the assumptions of Catholic inferiority (far more massively refuted by the evidence than assumptions of black inferiority) are never subjected to the slightest questioning? Why can America's intellectual and cultural elites so blithely assume that after a century and three-quarters nativism has finally been put to rest? Why indeed?

REFERENCES

1. Kenneth R. Hardy, "Social Origins of American Scientists and Scholars," *Science*, August 9, 1975, pp. 497–506.
2. Gerhard Lenski, *The Religious Factor* (Garden City, N.Y.: Doubleday, 1961), p. 255.
3. R. H. Knapp and H. B. Goodrich, *Origins of American Scientists*, 2 vols. (Chicago: University of Chicago Press, 1953).
4. Ibid., p. 46.
5. Ibid.
6. Ibid., p. 29.
7. Ibid., p. 48.
8. Thomas F. O'Dea, *American Catholic Dilemma: An Inquiry into the Intellectual Life* (New York: Sheed & Ward, 1958), p. 9.
9. Gustave Weigel, "American Catholic Intellectualism: A Theologian's Reflections," *Review of Politics* 19 (July, 1957): 299.
10. Karl F. Herzfeld, "Scientific Research and Religion," *Commonweal*, March 20, 1929, p. 562.
11. Weigel, "American Catholic Intellectualism," p. 299.
12. O'Dea, *American Catholic Dilemma*, p. 124.
13. Ibid., p. 142.
14. Daniel and Sidney Callahan, "Do Catholic Colleges Develop Initiative?" *Catholic World* 186 (December 1957): 182–183.
15. Joseph Cunneen, "Catholics and Education," *Commonweal*, August 7, 1953, p. 438.
16. John Tracy Ellis, "American Catholics and the Intellectual Life," *Thought* (Autumn 1955): 355–388.
17. Herzfeld, "Scientific Research," pp. 560–561.
18. Weigel, "American Catholic Intellectualism," pp. 289–290.
19. See Louis McKernan, C.S.P., "The New Religion of Science," *Catholic World* 86 (January 1958): 251; and Neil G. McCluskey, S.J., "Too Few Catholic Rhodes Scholars," *America*, April 7, 1956, pp. 26–30.
20. James A. Davis et al., *Stipends and Spouses: The Finances of American Arts and Science Graduate Students* (Chicago: University of Chicago Press, 1962).
21. Seymour Warkov and Andrew M. Greeley, "Parochial School Origins and Educational Achievement," *American Sociological Review* 31 (June 1966): 406–414.

22. Andrew M. Greeley, "Influence of the 'Religious Factor' on Career Plans and Occupational Values of College Graduates," *American Journal of Sociology* 6 (May 1963): 658–671.

23. Donald P. Warwick, "Letter to the Editor," *American Journal of Sociology* 69 (November 1963): 295.

24. Lenski, *Religious Factor*, p. 284, fn.

25. John Donaven, "Creating Anti-Intellectuals," *Commonweal* (October 2, 1964), pp. 37–39.

26. James Trent, *Catholics in College* (Chicago: University of Chicago Press, 1967).

27. Daniel Callahan, *The Mind of the Catholic Layman* (New York: Scribners, 1963).

28. Greeley, "Influence of the 'Religious Factor,'" p. 659.

29. Andrew M. Greeley, "Religion and Academic Career Plans: A Note on Progress," *American Journal of Sociology* 72 (May 1967): 668–672.

30. Andrew M. Greeley, "Continuities in Research on the 'Religious Factor,'" *American Journal of Sociology* 75 (November 1969): 355–359.

31. Andrew M. Greeley, *Religion and Career* (New York: Sheed & Ward, 1963).

32. Stephen Steinberg, *The Academic Melting Pot: Catholics and Jews in American Higher Education* (New York: McGraw-Hill, 1974).

33. Davis et al., *Stipends and Spouses.*

34. For a discussion of this finding, see Seymour Martin Lipset and Everett C. Ladd, "Jewish Academics in the United States: Their Achievements, Culture, and Politics," in *Jewish Year Book* ed. Michail Wallace (Hartford, Conn.: Prayer Book, 1971).

CHAPTER
5

Political Behavior

A PRINCIPAL THEME of American political analysis for the last decade has been realignment, the emergence of new political coalitions and new distributions of political power. Apparently this realignment process is continuing, and no definitive realignment has yet taken place to create a new majority party. As my colleagues Sidney Verba, John Petrocik, and Norman Nie pointed out, some realignment has taken place.[1] The South has become much more Republican and the Northeast much more Democratic than either section was previously. Blacks, already heavily Democratic, have become more so; blue-collar Protestants, once moderately Democratic, have become somewhat less Democratic. Independents are more numerous now than they have been for a long time, not so much because of a switch away from either of the traditional parties as because a whole new generation of politically unmobilized voters have come into adulthood in the last fifteen years. This younger generation is much less inclined than its elders to vote, and is as yet politically unaffiliated—very like the political unaffiliated of the 1920s. The "realignment" of 1932 occurred not so much because people stopped being Republicans and became Democrats (though in later years this apparently did happen with some black adults) but rather because the previously unmobilized younger generation of voters who grew up in the 1920s became Democratic in the Roosevelt election. Whether such a mobilizing election will occur in the next decade and perform a similar function for the political unaffiliates of the present era remains to be seen. There is a potential for realignment in this younger generation, and it does not seem to be leaning toward the Republicans. Thus it appears that at this time the realigning moves that have occurred tend to cancel one another out; the Democrats continue to be what they were since 1932, the majority party. Barring the emergence of a third party that would mobilize much of the young vote or a dramatic shift of this unaffiliated vote to the Republican side, the Democrats will continue to be the larger of the two political parties.

But before Verba, Nie, and Petrocik produced their careful analysis, a number of observers such as Kevin Phillips,[2] William Gavin,[3] Patrick Buchanan,[4] and William Rusher[5] described a "new Republican majority" which would include the economic conservatives (the old right wing of the Republican party) plus the social conservatives who were thought to be for the most part "the Catholic ethnics" who object to the "far out" liberalism of the left wing of the Democratic party. Those Catholic ethnics, in other words, who were lost to the Democrats in the McGovern election were viewed as potential components of a new Republican coalition. This feeling has been echoed just as consistently from the Democratic left, which has always been embarrassed by its Catholic ethnic allies, and particularly by the big-city urban "bosses" like Richard Daley and union "bosses" like George Meany. The intellectuals who designed the grand strategy of the McGovern movement—Galbraith, Burns, Dutton—thought that the Catholic ethnics could be conceded to the Republican party "where they belonged," and their departure could be made up for by the "legions of the young," who had come to voting age during the troubled Vietnam years. Unfortunately for their grand design, the majority of those under thirty were not like the students these worthy professors saw in their classrooms; like their parents, they were more inclined to vote for Nixon than for McGovern. Even among those under twenty-five, McGovern squeezed out a bare plurality. He did quite well with the college young, but the majority of American young people (55 percent) do not go to college, and in that group Nixon won overwhelmingly).

Since the late 1960s, then, there has been an assumption among right-wing Republicans that the Catholic ethnics "belonged" with them, and an assumption among Democrats that they did not need the Catholic ethnics. Both assumptions seem to have been wrong thus far. The McGovern movement went down to Goldwater-like defeat—in part because it lost substantially among Catholic voters; but the Republican party has not been able to capture any substantial support among Catholics and continues to be a party of about one-quarter of the country—a very homogenous one-quarter, predominantly white middle-class Protestant.

It does not seem to have occurred to writers like Rusher, Phillips, or Gavin that they may have misunderstood the conservative bent of urban Catholics, and that there may be considerably more distance between the so-called social conservatives and the economic conservatives than these architects of a new conservative coalition are inclined to believe. Similarly, it has yet to occur to many liberal left Democratic ideologues that Catholic ethnics, along with union members, Catholic and Protestant, are still an indispensable partner in any liberal coalition. Nevertheless, the evidence to be assembled in this chapter will indicate that the so-called social conservatives are much further away (both socially and economically) from the economic conservatives than either the conservatives or the liberals are prepared to believe.

Catholics have voted Democratic since the nineteenth century, when the Democratic party was able to make alliance with the new immigrants in the face of the nativist hostility of much of the Republican party. Despite ups and downs in this alliance (and the most severe down came at the end of World War I, when Woodrow Wilson appeared to many Catholics to be violently anti-Catholic), the urban ethnics and the Democratic party have been firmly aligned. Such issues as Prohibition strengthened the alliance, and, as the latest immigration of eastern and southern Europeans came of voting age in the 1920s, they inclined toward the Democratic party, first on the Prohibition issue and then because of the Al Smith candidacy. The Roosevelt New Deal in 1932 strengthened the alliance between Catholics and the Democratic party, and from 1932 to 1972 the Democratic candidates could always count on a minimum of three-fifths of the Catholic vote in the presidential elections. (The proportion fell just slightly below that even in the Republican landslide election of 1956.)

However, it has been argued by both the left, which wants the Catholics out of the Democratic party, and the right, which wants Catholics in a new Republican conservative coalition, that Catholics have been drifting away from the Democratic party because of their immigration to the suburbs and because of the increased importance of the "social" issue (which would include such issues as race, marijuana, amnesty for Vietnam draft dodgers, and abortion). Catholics were still voting Democratic, it might be admitted, but they were less likely to do so now. They were less likely to define themselves as Democratic, and their political attitudes were drifting to the right as the Democratic party drifted to the left. However, the data assembled in this chapter provide little support for such a view. Catholics have not left the Democratic party in affiliation; they are still substantially more likely than other Americans to vote Democratic in both presidential and congressional elections, and their 1972 defection was not disproportionate to that of other Democrats; their political and social attitudes put them on the liberal side of the American political spectrum.

A good deal depends on how you define "liberal" and "conservative," of course. Most Americans are not too consistent in their political and social attitudes; one achieves only a fair degree of predictive success if one tries to estimate a given person's attitude on, let us say, foreign policy from his attitude on, let us say, welfare reform. It is true, as Verba, Nie, and Petrocik have demonstrated, that political consistency has increased considerably in the last two decades (apparently as part of one of the great flings upward and downward in consistency that occur periodically in American history). But still, the overwhelming majority of Americans are neither consistently liberal nor consistently conservative on all salient political issues. If to be a "liberal" one has to hold the official liberal position on every issue, then not very many people in this society are liberal; the majority of American Catholics—unlike the majority of Jews and blacks—would not fall on the liberal *end* of the political spectrum. Furthermore, if one has to endorse the counterculture style and its "new morality," then not only most Catholics, but also most Jews and blacks, are not liberal. However, if liberalism means support for the New Deal "bread and butter" social stance, then the overwhelming majority of Catholics are still politically liberal. On almost all the issues that are pertinent in American political life today, Catholics are more likely than Protestants to fall on the liberal side of the political spectrum.

Let us begin with the issue of party affiliation. Since the work of Scott Greer[6] on the St. Louis metropolitan area, it is generally assumed that as the Catholic ethnics make money and move to the suburbs, they tend to weaken their ties with the Democratic party. Such ties are considered to be the product of both poverty and of the tight constraints of intercity ethnic neighborhoods. It occurred to very few observers, however, that St. Louis, with its strong German population, might not be the typical American urban setting. Using data collected in the University of Michigan's national election studies over the past two decades, my colleague Joan Fee[7] attempted a complex and sophisticated analysis of the question of the possible disaffiliation from the Democratic party of Catholics as they became better educated and more suburbanized. Ms. Fee deals with voters at three points in time, an early period (1952, 1956, 1958), a middle period (1960, 1962, and 1964), and a late period (1968, 1970, 1972). Each period contained two presidential elections and one congressional election. In time one, 42 percent of the Catholic population lived in the suburbs; by time three, 68 percent of the Catholics lived in the suburbs; but there was virtually no change in the proportion of Catholics identifying with the Democratic party. There was a decline of party identification because younger Catholics, like all other younger people, are more likely to be independent; but among party affiliators, there was little Catholic erosion from the Democratic party in the two decades. There was virtually no difference in Democratic party affiliation between suburban Democrats and those still living in the central city.

Analyzing the same national election file used by Ms. Fee, John

Petrocik[8] found that on a scale made up of six political opinion items (welfare, race, Cold War, school integration, and size of government), Catholics were on the average slightly left of center in the late 1950s and still left of center at the end of the 1960s—although there had been a slight movement to the right during the two decades. However, the proportion of Catholics at the far liberal end of the political continuum has actually increased during the period Petrocik studied.

TABLE 5.1

Where Ethnics Stand on Voting Patterns and Party Affiliation

	IRISH CATHOLICS	GERMAN CATHOLICS	ITALIAN CATHOLICS	POLISH CATHOLICS	ALL AMERICANS
Average percent voted Democratic Congress, 1952-1970	70	57	62	76	55
Average percent voted Democratic, 1970	73	45	62	73	54
Average percent voted Democratic for president, 1952-1968	65	55	60	76	48
Average percent voted for Humphrey, 1968	65	55	59	80	44
Percent Democrat or Democrat-leaning, spring 1972	61	57	63	72	55

SOURCE: Andrew M. Greeley, *Ethnicity in the United States: A Preliminary Reconnaissance*, p. 210. (This and following tables and figures from this source are reprinted by permission of the publisher, John Wiley & Sons — © 1974.)

When we turn from affiliation to voting, one can see in Table 5.1 that the four major Catholic ethnic groups voted in overwhelming numbers for Democratic congressional candidates between 1952 and 1970. With the exception of the German Catholics, there was virtually no difference in the 1970 Democratic congressional vote from the two-decade average. Furthermore, there was virtually no difference for each of the ethnic groups in the proportion voting for Hubert Humphrey when compared with the average proportion voting Democratic in the presidential elections of the 1950s and 1960s. The Poles, who were supposed to have voted for Wallace in large numbers, were even more likely to vote for Humphrey (80 percent) than their two-decade Democratic presidential average (which was still a very high 76 percent). Each of the Catholic ethnic groups was more likely to vote for George McGovern in 1972 than was the national American average. And the Gallup poll showed that during the Watergate crisis Catholics turned against Nixon earlier and more strongly than did other Americans.

It was during the late 1960s and the early 1970s that the image of the white ethnic hardhat racist polluter hawk seemed to grip the editorial writers, columnists, and feature story writers in the national journals. However, the national survey data of the time offered almost no confirmation for this image. In Table 5.2 one can compare the Italian and Polish ethnics with the Anglo-Saxon Protestants. The order of importance of the first five issues listed is precisely the same: war, marijuana, inflation, crime, and pollution. However, the Italians and the Poles were more likely than the Anglo-Saxon Protestants to see each of these as very serious problems. More than half of both groups thought that inflation, crime, and pollution were very serious problems, and three-fifths of the Italians and three-fifths of the Poles thought that marijuana was a very serious problem, putting it in second place on their list, as did Anglo-Saxon Protestants.

Table 5.3 gives the proportions of respondents in each group who both liked and disliked various elements in the society. The overwhelming majority of all the groups were sympathetically disposed to college students, indicating that the younger generation is by no means despised by most Americans. The Irish, Italians, and Poles are somewhat less likely, however, to like college students than are Anglo-Saxon Protestants. Similarly, while the majority of all the groups like college professors, the Irish and the Italians are somewhat less inclined to like them, and only a minority of the Poles (40 percent) are willing to admit that they like college professors. But while students and professors get good marks from the ethnics, radical students get very poor marks, with more than four-fifths of each group (and more than four-fifths of the Anglo-Saxon Protestants concur) registering dislike of radical students.

Admiration for big business is lower among the Catholic ethnics than it is among Anglo-Saxon Protestants, although only the Italians are notably more inclined to like labor unions. Liberals enjoy a better image with the Irish than they do with the other five groups, though only the Italians are less sympathetic to the liberals than are Anglo-Saxon Protestants. Similarly, the German, Italian, and Polish Catholics are somewhat less likely than are Anglo-Saxon Protestants to have a favorable image of conservatives.

There is little difference between the Anglo-Saxons and the Germans, Italians, and Poles in their images of blacks. The Irish, however, are notably more inclined to have a favorable image of them. All groups have a very unfavorable image of black militants. The Irish, German, and Italians have about the same image of welfare recipients as do the Anglo-Saxon Protestants, although the Polish image is distinctly unfavorable. Finally, all five groups are very favorably disposed to the police.

The ethnics are not "pure" liberals, and indeed a substantial propor-

TABLE 5.2

Where Ethnics Stand on Rank Order of First Six "Very Serious Problems," NORC 1971

RANK OF PROBLEM	ANGLO-SAXON PROTESTANTS		IRISH CATHOLICS		GERMAN CATHOLICS		ITALIAN CATHOLICS		POLISH CATHOLICS	
	PROBLEM	PERCENT	PROBLEM	PERCENT	PROBLEM	PERCENT	PROBLEM	PERCENT	PROBLEM	PERCENT
1.	War	68	War	69	War	62	War	73	War	71
2.	Marijuana	51	Inflation	55	Inflation	49	Marijuana	60	Marijuana	60
3.	Inflation	48	Crime	52	Pollution	46	Inflation	57	Inflation	57
4.	Crime	44	Welfare	42	Marijuana	42	Crime	55	Crime	55
5.	Pollution	42	Unemployment	38	Crime	39	Pollution	54	Pollution	55
6.	Unemployment	30	Pollution	38	Unemployment	30	Welfare	53	Urban unrest	35

SOURCE: Greeley, *Ethnicity in the United States*, p. 212.

TABLE 5.3

Where Ethnics Stand on Their Likes/Dislikes

OBJECT OF LIKES/DISLIKES	ANGLO-SAXON PROTESTANTS	IRISH CATHOLICS	GERMAN CATHOLICS	ITALIAN CATHOLICS	POLISH CATHOLICS
College students	80/3	67/0	80/2	70/0	72/6
Big business	76/14	42/21	31/26	57/10	33/21
Liberals	27/27	39/28	31/26	15/25	28/21
Blacks	50/12	79/0	49/7	50/2	46/13
Conservatives	46/12	46/18	30/13	38/8	31/13
Hippies	7/77	14/58	3/70	3/66	2/79
Radical students	4/86	7/86	2/81	14/82	0/88
Welfare recipients	30/21	28/17	33/28	28/18	10/35
College professors	63/7	57/10	67/3	57/7	40/2
Police	87/3	86/0	91/5	90/0	78/4
Unions	43/33	43/21	46/30	50/19	45/20
Black militants*	2/83	0/96	2/83	5/78	4/86

SOURCE: Greeley, *Ethnicity in the United States*, p. 213.

*The black response to this question is 35/37.

tion of them seem to dislike "pure" liberalism. Table 5.4 suggests that there is a strong reserve of potential support among the ethnics for liberalism. The majority of the Irish, for example, favored withdrawal from Vietnam, stiff punishment for pollution, a speeding up of racial progress, government support for family assistance, and a maximum use of resources to eliminate poverty. On each of these issues the Irish are more "liberal" than Anglo-Saxon Protestants. A surprisingly large minority of the Irish (30 percent) were also willing to support the legalization of marijuana (perhaps because the Irish romance with John Barleycorn makes them reluctant to deny any kind of narcotic to others). The Irish are also considerably more likely than the Anglo-Saxon Protestants to support government pressure for neighborhood integration and to oppose punishment for student rioters, although in both cases only a minority of the Irish respondents took such a position. On the other hand, 51 percent of the Irish respondents favored the use of force to end urban unrest. Finally, the Irish are the most likely of all the groups under consideration to think that the courts have gone too far in protecting the rights of criminals.

A majority of the German Catholics take the "liberal" side only on stiff penalties for pollution, though Germans are more likely than Anglo-Saxon Protestants to be on the liberal side on the war, on solving the underlying problems that cause riots, on family assistance, marijuana, and neighborhood integration.

Italian Catholics provide majority support for the liberal response on the war, pollution, family assistance, and government efforts to

TABLE 5.4

Where Ethnics Stand on Solutions to Issues, NORC 1971

ISSUE	ANGLO-SAXON PROTESTANTS	IRISH CATHOLICS	GERMAN CATHOLICS	ITALIAN CATHOLICS	POLISH CATHOLICS
Vietnam: withdrawal/victory	40/36	64/32	46/28	53/32	41/30
Pollution: fines and jail/nothing	71/19	77/23	85/4	83/4	79/11
Riots: solve problems/force	47/38	35/51	58/28	40/52	59/29
Racial progress: speed up/slow down	40/32	57/19	32/27	43/40	22/35
Family assistance: government support/support self	37/40	42/24	42/25	55/30	30/65
Marijuana: allowed/not allowed	8/87	30/52	21/69	12/88	13/79
Rights of criminals: more careful/too far	27/58	13/73	19/64	33/59	17/68
Forced neighborhood integration: government should/should not	22/67	34/49	27/55	28/55	22/23
Student rioters: should not punish/punish severely	11/74	20/67	8/74	23/59	10/75
Government to eliminate poverty: all resources/has done too much	43/37	66/17	45/31	62/31	74/11

SOURCE: Greeley, *Ethnicity in the United States*, p. 214.

eliminate poverty; they are also more concerned than any of the other groups for the protection of the rights of criminals and the nonpunishment of student rioters.

Finally, even though the Poles were the least likely of the ethnic groups to espouse a "liberal" position, 79 percent of them favored strong punishment for pollution, and 74 percent (highest of any of the five groups) favored all possible government resources used to eliminate poverty.

Using data collected at about the same time, Professor James Wright observed in an article in *The Nation*[9] that Catholics were less likely than Protestants to own guns. Among non-South whites living in cities or suburbs, half the Protestants own a gun and one-quarter own a pistol, while only 25 percent of the Catholics own a gun and 10 percent own a pistol. More than 90 percent of the Catholics who do not own guns support gun control, as do more than three-fifths of the gun owners. Two-thirds of the British Protestant population supports gun control, as does 85 percent of the Irish Catholic, 72 percent of the German Catholic, 81 percent of the Italian, and 74 percent of the Polish Catholic groups. Thus the image of the armed and dangerous white ethnic, ready to shoot on sight any hippie or black who threatens his neighborhood, may have a mythological utility to certain other segments of American society, but it scarcely corresponds with the facts.

During the same late 1960-early 1970 period, the so-called "blue-collar ethnic" supported the eighteen-year-old vote by a two-thirds majority, supported basing welfare on the cost of living by more than a three-quarters majority, and by more than half was deeply concerned about pollution. Three-fourths would vote for a black president, and more than half rejected the proposition that integration was moving too fast. Almost half of this same population would favor a guaranteed annual wage—almost 15 percentage points higher than the national average (Table 5.5).

It is interesting to note that blue-collar workers, including Protestants, were more likely than the national average to favor withdrawal from Vietnam within eighteen months in February 1970. It is also worth noting that white-collar Catholics were substantially more likely to be in favor of withdrawal than were white-collar Protestants, and that Catholic professionals and executives were the most likely of any group shown in Table 5.6 to be opposed to the war.

The data in Table 5.6 were collected in 1970, when it was permissible and even fashionable to be against the war. I was told when I cited this datum that the ethnics were recent converts to the cause of peace. However, on the dove scale constructed by Sidney Verba and his colleagues at Stanford in their 1967 study of attitudes on the Vietnam

TABLE 5.5

Attitudes on Certain Political Issues

	NORTHERN WHITE BLUE-COLLAR WORKERS		
ATTITUDE	PROTESTANT	CATHOLIC	NATIONAL AVERAGE
In favor of 18-year-old vote			
1968	62% (217)	65% (150)	66%
1969	62% (221)	64% (129)	61%
In favor of basing welfare on cost			
of living	79% (191)	79% (136)	77%
In favor of guaranteed annual wage	29% (228)	47% (134)	32%
Would vote for a black president	72% (193)	78% (142)	67%
Integration moving too fast	57% (280)	42% (127)	48%
Deeply concerned about pollution	50% (258)	55% (110)	51%

SOURCE: Andrew M. Greeley, "Political Attitudes of American Ethnics," *Public Opinion Quarterly* 36 (Summer 1972): 216. (This and following tables from this source are reprinted by permission of the publisher, Columbia University Press — © 1972.)

war (Table 5.7),[10] they were substantially more likely to be at the dove end of the scale than were native American Protestants. Only the eastern European Catholics were lower on the dove scale than native American Protestants. In other words, three of the four ethnic groups were not only *not* more hawkish in 1967, they were less hawkish than native Americans. I wonder why so few of the leaders of the peace movements were aware of white ethnic antipathy toward the war, and why so many were eager to write off white ethnics as hardhat hawks. The ethnics cannot be expected to hate a society which has made possible for them the kind of life their parents, grandparents, and great grandparents would never have dreamed possible. A position toward the war that might have been effective among them would have appealed to patriotism instead of ridiculing it.

In a 1973 political attitude study carried out by professors Nie, Verba, and Petrocik, evidence was found to confirm and complete this

TABLE 5.6

Attitudes Toward the War, February 1970

(percent in favor of immediate withdrawal or within 18 months)

	NORTHERN WHITES	
OCCUPATIONAL GROUP	PROTESTANTS	CATHOLICS
Blue collar	50% (195)	47% (114)
Clerical, sales, service	32% (73)	38% (60)
Professional and executive	39% (160)	52% (101)
National average	46%	

SOURCE: Greeley, "Political Attitudes of American Ethnics," p. 215.

TABLE 5.7

*Attitudes of American White Ethnic Groups
Toward the Vietnam War, 1967*

GROUP	PERCENT DOVE
Jews	48% (47)
Western European Catholics	29% (149)
Southern European Catholics	26% (60)
Western European Protestants	17% (499)
Native Americans	15% (104)
Eastern European Catholics	7% (42)

SOURCE: Greeley, "Political Attitudes of American Ethnics," p. 215.

picture of Catholics as moderately liberal on "New Deal" social issues.[11]

Catholics, whether they are Democrats, independents, or Republicans, are generally more likely to support the traditional New Deal social reforms—government intervention, government support of employment, income equalization, and medicare—than are their Protestant counterparts. Indeed, there is substantially less correlation on the first two items between affiliation and attitude for Catholics than there is for Protestants. The proportion of Catholic support declines only slightly as one moves from right to left across Table 5.8. There is, on the other hand, generally relatively little difference between northern Catholics and northern Protestant independents and Republicans on such civil liberties issues as marijuana, pornography, strictness against criminals, and spying on radicals.

On racial issues such as black welfare and compulsory school and housing integration, the majority of Americans do not side with the liberal coalition, but Catholic independents are more likely than their northern Protestant counterparts to support promotion of black welfare and compulsory school integration. They are also more likely to support such programs than are Catholic Democrats. Indeed, among white Gentiles, Catholic independents have the highest proportion of support for both the welfare and the integration questions (Table 5.8, part C).

Elections will still be won or lost in the United States on the traditional New Deal issues. Marijuana and pornography are matters that will be worked out by the state legislatures and the courts. The American public favors equality of opportunity, but it does not like quotas, reverse discrimination, compensatory justice, affirmative action, or whatever else such policy might be called. But most Americans are not personally affected by either busing or quotas, and they will not, I suspect, make such issues decisive in elections unless and until they are personally affected by them. In any event, one can conclude from Table 5.8 that the liberal coalition which is currently dominant in the struc-

TABLE 5.8

Attitudes on Party Affiliation and Population Group for 1973 on Traditional "New Deal," "Civil Liberties, and Racial Issues—% "Liberal"

A. TRADITIONAL "NEW DEAL" ISSUES

SOCIO-ETHNIC GROUP	JOBS			GOVERNMENT INTERVENTION			INCOME EQUALIZATION			MEDICARE		
	DEMO-CRATIC	INDEPEN-DENT	REPUB-LICAN	DEMO-CRATIC	INDEPEN-DENT	REPUB-LICAN	DEMO-CRATIC	INDEPEN-DENT	REPUB-LICAN	DEMO-CRATIC	INDEPEN-DENT	REPUB-LICAN
White northern Protestants	50	38	31	38	35	22	64	63	40	72	84	73
White border state Protestants	52	28	22	32	14	24	64	28	48	75	71	57
White southern Protestants	47	30	25	38	20	23	52	49	41	75	79	61
Catholics	58	50	49	46	37	39	63	67	54	84	79	66
Jews	55	—*	—	60	—	—	72	—	—	95	—	—
Blacks	86	—	—	68	—	—	86	—	—	92	—	—

B. CIVIL LIBERTIES

SOCIO-ETHNIC GROUP	MARIJUANA			PORNOGRAPHY			STRICT ON CRIMINALS			SPY ON RADICALS		
	DEMO-CRATIC	INDEPEN-DENT	REPUB-LICAN	DEMO-CRATIC	INDEPEN-DENT	REPUB-LICAN	DEMO-CRATIC	INDEPEN-DENT	REPUB-LICAN	DEMO-CRATIC	INDEPEN-DENT	REPUB-LICAN
White northern Protestants	35	32	17	40	41	31	27	26	19	54	51	38
White border state Protestants	7	14	18	30	14	23	43	14	27	47	43	28
White southern Protestants	10	13	20	26	47	19	26	20	27	43	39	39
Catholics	28	30	20	31	43	35	33	23	24	48	46	49
Jews	72	—	—	70	—	—	55	—	—	92	—	—
Blacks	44	—	—	41	—	—	52	—	—	59	—	—

C. RACIAL ISSUES

SOCIO-ETHNIC GROUP	BLACK WELFARE			SCHOOL INTEGRATION			HOUSING INTEGRATION		
	DEMO-CRATIC	INDEPEN-DENT	REPUB-LICAN	DEMO-CRATIC	INDEPEN-DENT	REPUB-LICAN	DEMO-CRATIC	INDEPEN-DENT	REPUB-LICAN
White northern Protestants	27	18	17	34	17	27	35	34	18
White border state Protestants	35	29	16	42	29	37	34	43	24
White southern Protestants	14	17	18	18	13	23	17	13	19
Catholics	30	35	12	36	49	24	34	24	23
Jews	40	—	—	52	—	—	50	—	—
Blacks	80	—	—	76	—	—	82	—	—

*Insufficient Jewish and black independents and Republicans to give percentage.

ture of the Democratic party (made up mostly of blacks, Jews, and very high-status northern white Protesants) is likely to find that on both the traditional New Deal issues and on the newer racial issues, the strongest potential support will come from that group of independents whose religious background would predispose them to be part of the Democratic coalition.

To summarize the state of our knowledge, the Catholic marriage with the Democratic party has not come to an end. Catholics are not leaving the Democratic party, despite their notable increase in education and their greater propensity to live in the suburbs. Up to the 1972 presidential election, they consistently voted for Democratic presidential and congressional candidates by a 60 percent or better majority. Their defection from the Democratic camp in 1972 was transitory, and their political beliefs and practices have not grown notably more conservative in the last two decades. Catholics are about where they were twenty years ago in their political and social attitudes, somewhat to the left of center. They are logical partners in any liberal coalition. If they are compared with Jews, university professors at elite universities, and writers of the national media, as well as other such elite groups, they are conservative, but then so is 90 percent of the country. But if the liberal-left elite of the Democratic party is looking for a coalition partner with which to win elections (which is what politics is about—or at least used to be about), then there is simply no other group in American society to look to except Catholics. The power of mythology, of un-examined symbolic assumptions, of pictures deeply rooted in the uncon-scious that blind one to reality and turn potential allies into imaginary enemies, must be very great indeed to keep alive the fiction of the blue-collar conservative ethnic.

We have argued in previous chapters that the ethnics are, if any-thing, less blue-collar than the American average. Now we argue that they are less conservative than the American average. The evidence is, I think, so overwhelming as to be incontrovertible; yet there are many people who simply do not want to believe it, even though their accept-ance and utilization of it could assure their return to and continuation in political power for the rest of the century. Unconscious imagery and unexamined assumptions, as well as unacknowledged prejudice, can block the realization of what would appear to be the most attainable of goals, the election to power of majority-party candidates.

There is no evidence to sustain the hoary image of the right-wing, anti-civil-libertarian Catholic which so delighted the cultural and in-tellectual elites during the McCarthy era. After all, Senator Joseph McCarthy was Irish Catholic, and didn't it follow that at least a majority of his fellow Irish Catholics would support the senator's anti-communist crusade? In fact, the evidence available even at that time provided little consolation for that view. And a redo in the early 1970s by the NORC General Social Survey of the questions asked in the famous Stouffer Civil Study[12] twenty years before shows that even when one confines one's investigation to the metropolitan regions of the north (thus excluding large numbers of antilibertarians, rural and southern Protestants), Catholics are at least as libertarian as Protestants, and in most cases somewhat more so (Table 5.9).

Senator McCarthy's own ethnic group, the Irish Catholics, are the most likely of all to endorse the right of the socialist to speak, and they are first among all the Gentile groups on the speaking and teaching rights of atheists and communists, as well as the most likely of the Gentile groups to defend the right of a Communist book to appear in the public library. Their Italian coreligionists are generally the second most liberal of the Gentile groups, though they are ahead of their Irish counterparts in their support for books by atheists to appear in the public library—and here their support is even higher than that of Jewish respondents. The Polish and the Slavic Catholics generally are indistinguishable on civil libertarian issues from the mainstream "other" or "American" Protestant groups. This is despite the presumed hostility of these eastern European groups to the Communist regimes that dominate their mother countries.

Political scientists are interested in the whole range of political behaviors. Verba and Nie have developed a "political participation factor"[13] (Figure 5.1) composed of four component factors: voting (Figure 5.2), political activism (campaigning, contributions—Figure 5.3), communal participation (membership in civic organizations—Figure 5.4), and "particularized contact" (writing letters to or personally contacting political leaders—Figure 5.5). Irish Catholics are the highest on the political participation scale, and also on the activism scale; they tie with Scandinavian Protestants on the communal participation scale. Polish Catholics are the most likely to vote, followed by the Slavic Catholics, themselves followed by the German, Irish, and Italian Catholics, with the Protestant groups trailing behind them. In addition, the eastern and southern European groups are the most likely to engage in particularized contact, with the Italians, Slavs, and Poles being ahead of Irish and German Catholics and all Protestants in their propensity to this kind of political behavior. (Perhaps it is the sort of behavior that occurs in neighborhoods where the precinct captains and other minor political figures are relatively accessible.) When we hold social class background constant, these patterns of political participation do not change.

So some Catholic groups are more active than others; indeed, the Irish Catholics are the highest political participators in the country. But all Catholics have one dimension of political participation at which they are the most active: Poles in voting, and Poles, Italians, and other

TABLE 5.9

Civil Liberties Attitudes—NORC General Social Surveys 1972, 1973, 1974
(percent metropolitan north only)

ETHNIC GROUP	ATHEIST*			SOCIALIST†			COMMUNIST‡		
	SPEAK	TEACH	BOOK	SPEAK	TEACH	BOOK	SPEAK	TEACH	BOOK
Protestants									
British(218)	73	43	55	89	63	54	64	36	69
German(247)	65	40	56	78	56	60	60	40	64
Scandinavian(74)	69	42	55	80	60	54	57	37	62
Irish(102)	66	43	52	74	52	57	57	63	59
Other(316)	64	37	52	74	53	58	53	34	56
Catholics									
Irish(113)	81	57	52	89	70	53	76	50	71
German(86)	77	51	57	81	61	62	66	38	62
Polish(71)	65	41	44	77	52	46	49	33	55
Slavic(34)	67	41	50	73	64	49	53	41	55
Italian(146)	77	50	59	84	64	56	64	45	59
Jews	83	68	54	87	77	58	80	60	83

Atheist attitude question: There are always some people whose ideas are considered bad or dangerous by other people. For instance, somebody who is against all churches and religion. . . .
A. If such a person wanted to make a speech in your city (town, community) against churches and religion, should he be allowed to speak, or not?
B. Should such a person be allowed to teach in a college or university, or not?
C. If some people in your community suggested that a book he wrote against churches and religion should be taken out of your public library, would you favor removing this book, or not?

†*Socialist attitude question:* Or consider a person who favored government ownership of all the railroads and big industries.
A. If such a person wanted to make a speech in your community favoring government ownership of all the railroads and big industries, should he be allowed to speak, or not?
B. Should such a person be allowed to teach in a college or university, or not?
C. If some people in your community suggested a book he wrote favoring government ownership should be taken out of your public library, would you favor removing this book, or not?

‡*Communist attitude question:* Now, I should like to ask you some questions about a man who admits he is a Communist.
A. Suppose this admitted Communist wanted to make a speech in your community. Should he be allowed to speak, or not?
B. Suppose he is teaching in a college. Should he be fired, or not?
C. Suppose he wrote a book which is in your public library. Somebody in your community suggests that the book should be removed from the library. Would you favor removing it, or not?

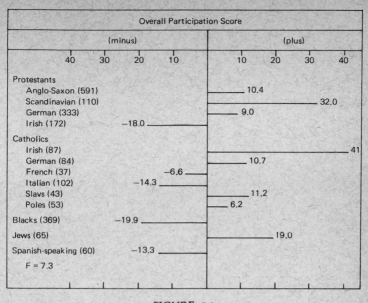

FIGURE 5.1

Political participation for American ethnic groups
(deviations from the mean).

SOURCE, figures 5.1–5.5: Greeley, *Ethnicity in the United States*, pp. 126–130.

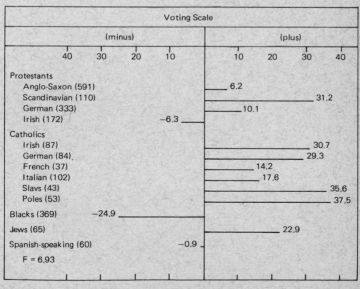

FIGURE 5.2

Political participation for American ethnic groups
(deviations from the mean).

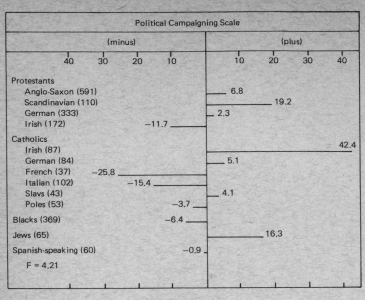

FIGURE 5.3

*Political participation for American ethnic groups
(deviations from the mean).*

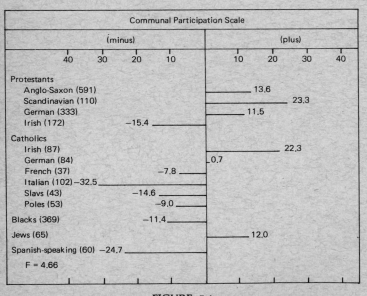

FIGURE 5.4

*Political participation for American ethnic groups
(deviations from the mean).*

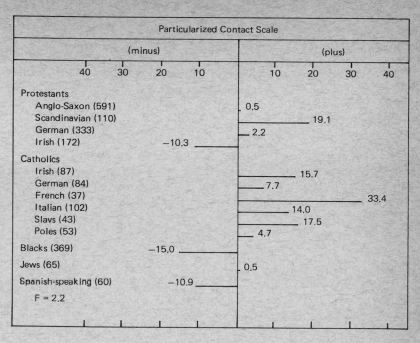

FIGURE 5.5

*Political participation for American ethnic groups
(deviations from the mean).*

eastern Europeans in particularized contact. Protestants lead in political participation of the communal or civic variety. Jews are also high on this form of participation, while Irish Catholics are the only Catholic group to score above the mean in civic involvement. Jews, for their part, are political activists and strong voters, but they are not much given to particularized contact. One sees the possibility for considerable misunderstanding of political styles across religious lines. The civic mode of participation is highly valued by Jews and Protestants but does not seem to attract Catholic groups besides the Irish. Particularized contact, on the other hand, is very popular with the eastern and southern European ethnic groups and not unpopular with the Irish, but it is not valued highly by either Jews or Protestants (except for German Protestants). Catholic groups (again excepting the Irish) seem to feel that they discharge their political responsibilities by voting and by particularized contact. Jews seem to feel that in addition to voting, one must also be involved in civic and campaigning activities. Protestants heavily emphasize civic participation. Responsible citizenship, then, means something different for Jews and Protestants than it does for eastern and southern European Catholics. The Italian and the Pole cast their votes, and when there is some kind of problem they go to their precinct captains. When the Protestant is faced with a problem, he joins a civic organization, and the Jew either joins a civic organization or engages in campaign activism. The Irish Catholic (hedging his bets, perhaps) engages in all four varieties of political participation. We have here, perhaps, a classic manifestation of the differences among ethnic subcultures which persist unaffected by education and generation in America and which fit neatly the permeable boundary-mosaic model presented in Chapter 1. Unless one understands that another group has a different political style and respects its right to that style (without necessarily being persuaded that it is the best style), one has the raw materials for both misunderstanding and conflict. Joining a civic group is not in itself either superior to or more effective than calling your precinct captain; neither is it necessarily inferior or less effective; it is simply different.

It may well be, however, that differences in political style are more important than differences of political substance. The misunderstanding and misperception of Catholic voting, political affiliation, and political attitudes may not be based so much on substantive differences between Catholics and other groups which are disproportionately represented in the Democratic party coalition as on subtle and unrecognized but powerfully important differences in the way with which one approaches political participation.* Catholics are not more conservative than other Americans, and they are certainly not more likely to have been hawkish on the Vietnam war or hardhat on the social issues. We shall see in Chapter 6 that they are not against integration. But they are more likely to call their precinct captain and less likely to join civic organizations, and that may be the real problem.

* Despte polls up to the Friday before the 1976 election showing that Jimmy Carter only had 45 percent of the Catholic voters in the country commited to him, it would seem that over the weekend he picked up virtually all the uncommitted Catholics, and in fact, gained 56 percent of the Catholic vote, including 65 percent of the blue collar Catholic vote. Despite considerable uncertainty and hesitation about Carter among Catholics, he still did substantially better than George McGovern (48 percent) though not as well as Hubert Humphrey (62 percent). It would appear that in an election in which neither candidate appeared terribly attractive to Catholic voters, the Democratic party preferences and past voting behavior was of decisive last minute importance to the undecided Catholics.

REFERENCES

1. Norman H. Nie, John R. Petrocik, Sidney Verba, *The Changing American Voter* (Cambridge, Mass.: Harvard University Press, 1976).
2. Kevin Phillips, *The Emerging Republican Majority* (Garden City, New York: Doubleday Anchor, 1969).
3. William F. Gavin, *Street Corner Conservative* (New York: Arlington House, 1975).
4. Patrick Buchanan, *The Conservative Choice* (New York: Quadrangle, 1975).
5. William A. Rusher, *The Making of a New Majority Party* (New York: Sheed & Ward, 1975).
6. Scott Greer, "Catholic Voters and the Democratic Party," *Public Opinion Quarterly* 25 (1961): 611-625.
7. Joan Fee, "Political Continuity and Change," in Andrew M. Greeley, William C. McCready, and Kathleen McCourt, *Cathole Schools in a Declining Church* (Kansas City: Sheed & Ward, 1976), pp. 76-102.
8. John R. Petrocik, "Changing Party Coalitions and the Attitudinal Basis of Alignment, 1952-1972 (Ph.D. diss., University of Chicago, 1975).
9. James Wright, "Who Owns the Side Arms" *The Nation*, September 20, 1975, pp. 240-244.
10. Sidney Verba et al., "Public Opinion and the War in Vietnam," *American Political Science Review* 61 (June 1967): 317-333.
11. Nie, Petrocik, and Verba, *The Changing American Voter.*
12. Samuel A. Stouffer, *Communists, Conformists, and Civil Liberties; A Cross Section of the Nation Speaks Its Mind* (Garden City, N.Y.: Doubleday, 1955).
13. Sidney Verba and Norman H. Nie, *Participation in America: Political Democracy and Social Equality* (New York: Harper & Row, 1972).

CHAPTER

6

Catholic Prejudice

IT IS generally assumed that Catholics and blacks are natural enemies. They are both struggling for the same neighborhoods and the same working-class jobs in the city. Jews and Protestants, at least in the North, are either less likely to be prejudiced or less likely to live in the cities and be in competition with blacks for the same jobs. It was the blue-collar ethnic racist who threw rocks at Martin Luther King in Marquette Park, Chicago; it was the blue-collar racist who fought busing in Michigan and who continues to resist it in South Boston.

In fact, as we have shown in previous chapters, Catholics are not at the bottom of the occupational scale in big cities, and indeed have become predominantly suburban in the last twenty years. Some Catholics still live in the city, and some are closely juxtaposed in neighborhoods and on the job with blacks; but the Irish in South Boston who oppose school busing are not necessarily any more typical than the Irish judge who lives in Newton and imposes it on his more unfortunate confreres.

NORC has been monitoring racial attitudes since 1940. In a series of articles written by Paul Sheatsley and Herbert Hyman (surveying the years 1940–1964) and Paul Sheatsley and Andrew M. Greeley (for the years since 1964),[1] changes in attitudes over time were carefully documented. Northern white Catholics, at every point in time, were less likely than northern white Protestants to be opposed to racial integration. In analyzing data collected in a 1968 survey, Normal Nie, Barbara Currie, and I looked at each of the five items in the NORC Integration Scale and discovered that on all items the Catholic ethnics, including the southern and eastern Europeans, were more inclined (or at least as likely) to be prointegration than their Protestant counterparts.[2] On the summary measure of the five items, the least prointegration of the Catholics, the eastern Europeans, were still substantially ahead of their Protestant counterparts.

Region appears to account for much of the difference in racial attitudes between ethnics and nonethnics. The total proportions of each of the ethnoreligious groups giving responses favorable to blacks become nearly identical when the South is excluded from the scale. A comparison of the right-most column of percentages in the race breakdown portion of Table 6.2 with the similar breakdown in Table 6.1 (which includes the South) tells us clearly why this is the case. The proportion of each of the *ethnic* groups favorable to blacks changes very little, primarily because such a small fraction of each of these groups lives in the South. However, there is a significant shift upward in the proportion of the native Protestants who report favorable attitudes toward blacks when the southern component of this group is excluded.

The data in Table 6.2 suggest that when it comes to racial attitudes, ethnic background appears to matter very little for those who live outside the South and earn more than $5,000 a year. However, among those who earn less than that the situation is different. Low-income ethnic groups are considerably more problack on our measure than are either their native Protestant counterparts or their wealthier ethnic cousins.

So it is precisely among the eastern and southern Europeans who do not earn high incomes—the blue-collar ethnics—that the differences between Catholics and Protestants outside the South are most striking. These people, who have been stereotyped as bigots and who have been made to pay the costs of racial justice in their schools, neighborhoods, and jobs, are the least likely to oppose racial integration. The person who acquires his view of American racial attitudes from watching television has a clear and vibrant image of the Polish truck driver screaming racist epithets at the TV camera. Of course, assumes the viewer, such a "neighborhood leader" is typical of the neighborhood. Such a line of reasoning may be legitimate among TV journalists and encouraged by their reporting; it is scarcely legitimate among educated persons who take seriously systematic research.

Curiously, it is also the eastern and southern Europeans who are the most likely to live in racially integrated neighborhoods. In areas outside the South, 18 percent of the eastern and southern European Catholics live in racially integrated neighborhoods, as opposed to 6 percent of the Protestants. But it is in such neighborhoods that one finds the biggest difference between southern and eastern European Catholics and white Protestants in their attitudes toward blacks (Table 6.3). Thus, whatever the TV commentators may say about Marquette Park in Chicago and South Boston, there is some reason to suspect that white

TABLE 6.1

Racial Attitudes among Ethnic Groups from the NORC Integrated Neighborhood Study (percent)

RELIGIO-ETHNIC GROUP	FAVOR SCHOOL INTEGRATION	WOULD BRING BLACK TO DINNER	OPPOSE KEEPING BLACKS OUT OF NEIGHBORHOOD	OPPOSE LAWS AGAINST RACIAL INTERMARRIAGE	DISAGREE THAT BLACKS SHOULD NOT PUSH	SUMMARY SCALE*
Western European Protestants (783)†	70	59	45	42	13	25
Irish Catholics (56)	82	79	68	61	20	43
German Catholics (47)	79	68	60	57	17	40
Southern European Catholics (39)	80	69	54	56	13	39
East European Catholics (54)	78	57	43	59	13	35
Jews (24)	100	92	79	75	33	67

SOURCE: Greeley, *Ethnicity in the United States*, p. 194.

*Percent giving four or five problack answers.

†Numbers of cases on which the percentages are based vary slightly from item to item depending upon missing data.

TABLE 6.2

Political Attitudes of Ethnics and Nonethnics by Family Income Outside the South

RELIGIO-ETHNIC GROUP	LESS THAN $5,000	$5,000-9,000	$9,000+	TOTAL NON-SOUTH
Race scale (percent problack)				
Native Protestants	66 (250)	66 (325)	63 (325)	65
Western European Catholics	76 (49)	68 (85)	66 (110)	69
Southern/Eastern Europeans	83 (56)	60 (70)	60 (65)	67
Jews	— (6)*	57 (14)	64 (30)	62
Service scale (percent favoring government help)				
Native Protestants	76	57	50	60
Western European Catholics	75	71	57	65
Southern/Eastern Europeans	82	80	72	78
Jews	—	86	71	76
Help poor (percent favoring government help)				
Native Protestants	68	53	49	56
Western European Catholics	67	66	51	59
Southern/Eastern Europeans	74	69	62	68
Jews	—	57	64	64
Hawk-dove scale (percent dove)				
Native Protestants	49 (99)†	38 (151)	40 (176)	41
Western European Catholics	23 (21)	50 (47)	59 (68)	50
Southern/Eastern Europeans	46 (22)	39 (38)	66 (32)	50
Jews	— (2)	— (7)	94 (18)	—

SOURCE: Greeley, *Ethnicity in the United States*, p. 197.

*Too few cases for percentaging.

†Vietnam questions asked of only half the total sample.

ethnics are better able to cope with problems of racial integration than are their Protestant American counterparts.

The NORC racial integration scale was administered again in 1970 and 1972 (Table 6.4). Mean scores on the five-item scale were computed at the two different points in time so that one could measure where the Catholic ethnic groups stood in 1970, as that turbulent decade came to an end, and then see how their opinions on racial matters had changed during what was supposed to be a period of "racial backlash." In fact, support for racial integration increased among northern Americans in those two years (increasing even more sharply in the South). A sharp

TABLE 6.3

*Problack Attitudes of Ethnics and Nonethnics by Residence
in Racially Mixed Neighborhoods*

	PERCENT GIVING PROBLACK RESPONSES IN CENSUS TRACTS LESS THAN 5 PERCENT BLACK	PERCENT GIVING PROBLACK RESPONSES IN CENSUS TRACTS 5 PERCENT OR MORE BLACK
Native Protestants	65 (848)	62 (55)
Western Catholics	68 (235)	72 (20)
Southern/Eastern Europeans	64 (181)	77 (39)
Jews	66 (52)	83 (16)

SOURCE: Greeley, *Ethnicity in the United States*, p. 205.

increase occurred for the Irish Catholics in that 1970–1972 time period, making them second to the Jews in their prointegration attitudes. However, an even greater increase in support for racial integration occurred among Italian Catholics, whose score went up from 2.65 in 1970 to 3.14 in 1972. Italian Catholics were substantially below the northern average in sympathy for racial integration in 1970, but by 1972 they had caught up with the average. Only the eastern European Catholics score below the national average among the Catholic ethnic groups.

To the extent that there is a Catholic "problem" in the racial area, it seems to be limited to the eastern European Catholics, and then it seems to show up only on some surveys. However, even when they appear below the national average on racial prointegration attitudes,

TABLE 6.4

*Prointegration Scale by Ethnicity
(non-South only)*

	1970	1972	Change 1970-1972
All Northerners	2.88	3.16	.28
Anglo-Saxons	2.80(220)	3.18(148)	.38
German Protestants	2.81(137)	2.70(142)	−.11
Scandinavian Protestants	2.82(29)	2.98(65)	.16
Irish Catholics	3.06(48)	3.46(63)	.40
German Catholics	2.97(41)	3.18(44)	.21
Italian Catholics	2.65(38)	3.14(63)	.49
Slavic Catholics	2.45(53)	2.76(49)	.31
Jews	3.79(24)	3.67(52)	−.12

SOURCE: Greeley, *Ethnicity in the United States*, p. 219.

their progress toward the average is comparable with other northern groups.

A reanalysis of the 1968 study done by the University of Michigan of fifteen cities that had been troubled by riots provided some explanation for the negative reaction of southern and eastern European Catholics toward blacks on certain measures.* Jews and Irish Catholics were the most likely of any ethnic group studied in the fifteen cities to support a black mayor, to support civil rights legislation and interracial contacts, and to oppose repressive riot control. The Poles had the lowest scores on all these scales. An explanatory model was developed which took into account education and two "external" factors, "avoidance" and "support." The avoidance factor weighted heavily on items indicating few black friends and fear for one's neighborhood; the support factor weighted heavily on items showing perception of discrimination and sympathy for black protest (Table 6.5).

Table 6.6 shows that the explanatory model accounted for many of the differences between Poles and Jews and between Poles and Irish Catholics. These differences virtually vanish on support for a black mayor and support for repressive riot control. Only on matters of civil rights legislation and interracial contact do differences remain, but they are cut in half.

Why do the differences between Poles and Jews persist? There are two possible sets of explanations. It may be that there is something in the heritage with which a group comes to the United States that predisposes it to certain responses when faced with racial diversity. Or it may be that something has happened in the group's experience since it arrived here that has made it suspicious of other groups in the society. The real explanation is probably a complex interaction of the two. The Poles have a long history of oppression and betrayal, and of course there is nothing in Jewish history that would dispose that group to be very trusting either. The family structures of the two groups seem to differ. The Jews have a much stronger maternal figure in their family culture than do the Poles; perhaps those kinds of differences produce different responses to racial problems.

Polish immigrants did not come to the United States with a radical or socialist tradition like that held by many Jewish immigrants. Many Polish immigrants came from the farms and fields of southern Poland, while many of the Jewish immigrants were from more urban backgrounds. The differences between Poles and Jews in their attitudes toward race may have their roots in the peasant origins of one group

---

* The data were originally collected by Dr. Angus Campbell and Dr. Howard Schuman of the Survey Research Center, Institute for Social Research, University of Michigan. Neither the original collectors nor the archive bear any responsibility for the analyses or interpretations presented here.

TABLE 6.5

Reduction of Differences Between Jews and Poles and Poles and Irish Catholics by Explanatory Model (Standard Deviations)

	POLES AND JEWS	POLES AND IRISH CATHOLICS
A. Support for Black Mayor		
Raw difference	.50	.37
Standardized for education	.37	.32
Standardized for education and avoidance factor	.55	.21
Standardized for education and both factors	.11	.10
B. Support for Civil Rights Legislation		
Raw difference	.69	.29
Standardized for education	.59	.19
Standardized for education and avoidance factor	.53	.14
Standardized for education and both factors	.34	.00
C. Support for Repressive Riot Control		
Raw difference	.30	.30
Standardized for education	.26	.27
Standardized for education and avoidance factor	.14	.23
Standardized for education and both factors	.08	.16
D. Support for Interracial Contact		
Raw difference	.64	.40
Standardized for education	.51	.34
Standardized for education and avoidance factor	.47	.26
Standardized for education and both factors	.26	.15

SOURCE: Greeley, *Ethnicity in the United States*, p. 231.

TABLE 6.6

Z Scores for Certain Ethnic Groups on Racial Attitudes after the Effect of the Explanatory Model Has Been Taken into Account

RELIGIO-ETHNIC GROUP	VOTE FOR BLACK MAYOR	CIVIL RIGHTS LAWS	RIOT CONTROL*	INTERRACIAL CONTACT
British Protestants	−17	−04	00	03
Irish Catholics	07	04	−07	00
Italian Catholics	04	01	07	−02
Polish Catholics	03	03	10	−09
Slavic Catholics	25	−06	00	07
Jews	08	31	02	16

SOURCE: Greeley, *Ethnicity in the United States*, p. 232.

*High score indicates support for repression.

and the more urban origins of the other. However, both the Irish and Italian Catholics also have peasant origins, and they react very differently to racial questions.

The Jews have met with more success in the United States than have the Poles. Indeed, recent demographic research indicates that the Poles are lagging behind other Catholic ethnic groups in the pursuit of affluence. It may well be that as a group, Poles do not feel accepted in American society. They may not feel as upwardly mobile as others, and hence are more disposed to be suspicious of those beneath them than are groups like the Italians and the Irish, who are moving upward rapidly. Even a control for education would not eliminate such feelings. One may be educated and still feel one's group is not being given an equal chance. It is worth noting, incidentally, that the Slavic group (non-Polish eastern European Catholics) is very different from the Poles in its response to racial questions. It is clearly a challenge for future research to explore these differences and the reasons for them.

Do the Poles feel that somehow they have been left out of the American dream, or at least placed under severe limitations in achieving it? There were three measures in the riot study that might enable us to tap such feelings of alienation.[8] They are measures of "efficacy," "competence," and "trust" of government.* On all three of the scales the Poles scored substantially below the mean. When we correlated these alienation measures with racial attitudes for the Poles, we discovered that lack of trust in government correlates above .2 in the Polish subsample with attitudes on a black mayor and riot control; feelings of competence correlate above .2 with the attitudes on civil rights laws and riot control (Table 6.7). Thus it can safely be said that for the Poles there is some relationship between "alienation" and racial attitudes.

Data collected in the NORC General Social Survey of the early 1970s (Table 6.8) show that of all America's religio-ethnic groups the Irish were the most likely to say that they would vote for a qualified black president; approximately four-fifths of all Catholics would do so. Also, approximately half of all Catholics would accept an integrated school where more than half the students were black. (In 1958 fewer than half of the American Protestants said they would vote for a qualified Catholic for the presidency.) While most American Catholics are opposed to busing, so are most American Protestants and Jews. That support for integrated schools can coexist with opposition to busing may seem

* The efficacy scale was devised from a series of questions that tapped respondents' sense of control over their lives, their confidence in the future, and their ability to plan ahead. The competence scale was constructed from questions that showed respondents' confidence in their ability to get action from a local government official in response to complaints. Respondents' "trust in government" was determined through a series of questions that reflected confidence in local, state, and federal government officials and in their efforts to solve problems.

TABLE 6.7

*Correlations Between "Alienation" Variables and Racial
Response of Polish Catholic Religio-ethnic Group Only*

RACIAL ATTITUDES	TRUST	EFFICACY	COMPETENCE
Vote for black mayor	.20	.15	.16
Civil rights legislation	.08	.00	.21
Riot control	.31	.12	.23
Racial contact	.08	.02	.00
Support	.17	.07	−.06
Avoidance	.12	.12	−.12

SOURCE: Greeley, *Ethnicity in the United States*, p. 239.

inconsistent, but it obviously does not seem so to the overwhelming
majority of Americans.

Research carried on in the late 1960s and early 1970s, then, indicated that Irish Catholics were substantially higher in support for
integration and enlightened responses to urban unrest than were
Protestants from cities in the North. Italian Catholics were at or above
the average on racial attitude measures, and only Poles were beneath
the average on some but not all measures on some but not all studies.
The low support for racial integration among Poles seem to be the
result of their low perception of discrimination against blacks and fear
of racial integration in their neighborhoods. Both these fears are somehow linked to a sense of alienation among Poles, a feeling of being left

TABLE 6.8

*Racial Attitudes in 1970-1975—NORC General Social Surveys, Metropolitan North
(percent)*

ETHNIC GROUP	WOULD VOTE FOR BLACK PRESIDENT	INTEGRATED SCHOOL			FAVOR BUSING
		FEW	HALF	MORE	
Protestants					
British	70	97	74	46	7
German	81	96	70	44	12
Scandinavian	86	100	78	41	10
Irish	73	93	81	54	7
Other	73	93	75	53	14
Catholics					
Irish	87	97	72	52	7
German	81	95	71	53	19
Polish	79	98	72	46	7
Slavic	84	94	77	50	11
Italian	78	93	69	43	12
Jews	82	94	85	46	18

out or cut off from the rest of society. However, one survey showed that the eastern and southern Europeans who lived in racially integrated neighborhoods in the North were substantially more supportive of racial integration than white Protestants who lived in the same kind of neighborhood.

To the extent that there is any evidence for the stereotype of the Catholic ethnic racist in the late 1960s and early 1970s, it is limited to eastern European Catholics, and then it is by no means conclusive. The Irish Catholics were almost even with the Jews in the early 1970s in their support for racial integration.

Finally, one must investigate the possibility that Catholic education, which separates Catholics from other people in society, may be producing graduates who are more socially and racially intolerant. My colleague Shirley Saldanha investigated this possibility, using data from the 1963 and 1974 NORC parochial school studies.[4] She found there was a positive correlation between number of years in Catholic schools at both points in time and attitude toward blacks and Jews. The more Catholic education one had, the more enlightened one's sentiments toward minorities. And this was not merely a function of having had more education. Catholic schools made a contribution of favorable attitudes toward blacks and Jews above and beyond the contribution made by similar levels of educational attainment in other schools (Table 6.9).

Thus while there was little change in the racial attitude scale in the two parochial school studies (a scale not comparable to the NORC integration scale discussed previously), there were sharp declines on the racism scale for those who had eight to twelve years of Catholic education, and an even sharper decline for those who had thirteen or more years of Catholic education. Furthermore, the two highest groups in Catholic educational attainment were more likely to accept integrated education than those which had no Catholic education or less than eight years of it. Finally, when educational attainment was taken into account, the net correlation between years of Catholic schooling and racism score for all respondents declined from 0 to −7 between 1963 and 1974; for college-educated respondents, it declined from −.04 to −20. Since there is a −20 correlation between college attendance and racist attitudes, one can say that going to a Catholic college makes an independent contribution to racial attitudes above and beyond the sheer fact of college attendance. Going to Catholic college, in other words, doubles the antiracism payoff of college education for Catholics.

We have demonstrated thus far that as far as northern and southern European Catholics are concerned, the stereotype of the Catholic racist simply does not hold up. Indeed, Irish Catholics are the most racially enlightened of all the Gentile groups. We have not analyzed any data showing the attitudes of other groups toward Catholics, because very

TABLE 6.9
Catholic Education and Racial Attitudes—1963-1974

i. MEAN SCORE ON RACISM SCALE

YEARS OF CATHOLIC SCHOOLING	1963	1974
0	1.60	1.73
1-8	1.63	1.68
8-12	1.45	1.27
13 or more	1.31	0.94
All	1.57	1.60

ii. MEAN SCORE ON SUPPORT FOR INTEGRATED EDUCATION (PROPORTION OF BLACKS ACCEPTABLE IN SCHOOL)

0	1.98
1-8	1.55
8-12	2.02
13 or more	2.30

iii. NET CORRELATIONS (RELIGIOUS BEHAVIOR MODEL) BETWEEN YEARS OF CATHOLIC EDUCATION AND RACIAL ATTITUDES

	1963	1974	
	RACISM	RACISM	SUPPORT FOR INTEGRATION
All respondents	.00	−.17	.00
College educated	−.04	−.20	.11

SOURCE: Andrew M. Greeley, William C. McCready, and Kathleen McCourt, *Catholic Schools in a Declining Church* (Kansas City: ⌐heed & Ward, 1976), p. 193.

rarely have American survey researchers determined that anti-Catholic nativism was a matter worth investigating (with the exception of asking whether respondents would vote for a Catholic president—which item was dropped from the survey repertoire after 1960). The *Catholic Digest* surveys (carried out in 1953 by Ben Gaffin and Associates and in 1965 by Gallup)[5] on Catholic attitudes did provide some interesting evidence that in the decade between the early 1950s and 1960s there had been a change in mutual attitudes of Catholics and Jews. In the first point in time, Jews were more sympathetic to Catholics than Catholics were to Jews, but in the course of the decade Catholic attitudes toward Jews became more favorable, while Jewish attitudes toward Catholics became less so. The two groups passed each other in mutual regard during the late 1950s and early 1960s. By 1965 Catholics had more favorable attitudes toward Jews than Jews did toward Catholics, according to the Gaffin-Gallup data.

Saldanha also shows a significant relationship between attendance at Catholic schools—particularly at the college level—and low anti-Semitism scores, although there was no change in the size of this relationship between 1963 and 1974. In both cases, however, the influence of Catholic education was *in addition to* the influence of the level of education. Catholic education, in other words, made a significant and independent contribution to lower anti-Semitism scores among Catholics.

Since 1965, only one survey has provided data to enable us to explore this phenomenon, a survey of adolescents and their parents done by the University of Michigan in 1965 and repeated on the same sample eight years later, in 1973.[6] The item used from the questionnaire was a so-called "feeling thermometer" in which the respondents were asked to rate their feelings toward other groups on a scale running from zero to one hundred. It is perhaps a symptom of the troubled times the nation went through between 1965 and 1973 that of the twenty-four possible attitudes of white Protestants, Catholics, Jews, and blacks toward one another, only two had increased, and they were both in Protestant attitudes (toward Jews and toward Catholics) (Table 6.10). In 1965 there was no difference in the attitudes of Jewish adolescents toward Catholics and of Catholic adolescents toward Jews; both gave each other a score of sixty-nine on the zero to one hundred scale. In the ensuing decade the regard that both groups had for one another diminished—the Catholic attitude for Jews by six points and the Jewish attitude for

TABLE 6.10

Paired Comparisons of Attitudes in 1965 and 1973
(scores on a scale from 1-100)

	YOUNG PEOPLE			PARENTS		
	1965	1973	Difference	1965	1973	Difference
A. CATHOLICS AND JEWS						
Catholics toward Jews	69	63	− 6	73	69	−4
Jews toward Catholics	69	60	− 9	67	61	−6
B. CATHOLICS AND PROTESTANTS						
Catholics toward Protestants	79	68	−11	82	75	−7
Protestants toward Catholics	65	63	− 2	65	67	2
C. CATHOLICS AND BLACKS						
Catholics toward Blacks	71	62	− 9	67	65	−2
Blacks toward Catholics	66	60	− 6	69	62	−7

Catholics by nine points. Among adults in 1965 there was confirmation of the finding of the *Catholic Digest* study. Catholics gave Jews an average score of seventy-three on the feeling thermometer, while Jews gave Catholics a lower score of sixty-seven. There was a decline for both groups in the ensuing decade, so that the gap in mutual esteem between Catholics and Jews in the parental generation had widened by 1965, Jews giving Catholics a score of sixty-one and Catholics giving Jews a score of sixty-nine.

Similarly, at both points in time and in both generations, Catholics had substantially more favorable attitudes toward white Protestants than white Protestants did toward Catholics. At both points in time Catholic young people had more favorable attitudes toward blacks than black young people had for Catholics, although the differences had diminished between 1965 and 1973. In the parental generation, in 1963, blacks were slightly more favorable to Catholics than Catholics were toward blacks, but this situation had reversed itself by 1973.

In summary, the decline of good feeling among religio-racial groups in American society had produced a situation by 1973 in which Catholics of both generations had more positive attitudes toward Jews, white Protestants, and blacks than Jews, white Protestants, and blacks had toward them. These findings suggest at least tentatively that if there is a serious problem of intergroup relations persisting in American society, it is not so much Catholic ill feeling toward blacks, Jews, or white Protestants as it is ill feeling in these three groups against Catholics. Catholic anti-Semitism and Catholic racism have been the subject of much research and discussion, but anti-Catholicism has gone virtually unexamined.

With what must seem by now a monotonous regularity, the empirical evidence which is the raw material of this book has been at variance with the truth "everyone knows" about Catholics. Thus in the last two chapters we have seen that the political conservative "ethnic" turns out not to be conservative at all on the kinds of issues which win elections. The hawkish Catholic turns out to be dovish, the racist Catholic turns out to be less so than comparable groups from the general population. Catholics are less likely to be anti-Semitic than Jews are to be anti-Catholic. Attendance at Catholic schools correlates positively with tolerant attitudes on race and anti-Semitism. Anti-Catholic feeling appears to be on the increase, while Catholic antagonism toward other groups is not.

A friend of mine told me he had heard some of our findings discussed at a very proper Harvard dinner party. The verdict was that they were absurd and impossible.

That is, of course, one explanation.

REFERENCES

1. Herbert H. Hyman and Paul B. Sheatsley, "Attitudes Toward Desegregation," *Scientific American* 211 (July 1964): 2–9; Andrew M. Greeley and Paul B. Sheatsley, "Attitudes Toward Racial Integration," *Scientific American* 225 (December 1971): 13–19; and Andrew M. Greeley and Paul B. Sheatsley, "Attitudes Toward Racial Integration: The South 'Catches Up,' " in *Social Problems and Public Policy: I. Inequality and Justice,* ed. Lee Rainwater (Chicago: Aldine, 1974), pp. 241–250.

2. Norman H. Nie, Barbara Currie, and Andrew M. Greeley, "Political Attitudes Among American Ethnics: A Study of Perceptual Distortion," in Andrew M. Greeley, *Ethnicity in the United States* (New York: Wiley Interscience, 1974), pp. 186–216.

3. Andrew M. Greeley, "Ethnicity and Racial Attitudes: The Case for the Poles and the Jews," *American Journal of Sociology* 80 (January 1975): 909–933.

4. Shirley Saldanha, "The Social Attitudes of Catholics and the Influence of Catholic Education: A Comparison between 1963–64 and 1974," multilith (Chicago: NORC, 1976).

5. The *Catholic Digest* surveys were reported in John L. Thomas, S. J., *Religion and the American People* (Westminster, Md.: Newman Press, 1963).

6. See M. Kent Jennings, *The Student-Parent Socialization Study* (Ann Arbor: University of Michigan Inter-University Consortium for Political Research, 1971); and M. Kent Jennings and Richard G. Niemi, "The Transmission of Political Values from Parent to Child," *American Political Science Review* 62 (March 1968): 169–184.

CHAPTER

7

The Change in
the Church

ANYONE who reads the daily newspaper is aware that there has been major change in the Catholic church as institution. In the last decade the Vatican Council was front-page headlines: the mass has been changed into English, nuns wear secular dresses, priests and nuns have resigned from their work in substantial numbers, church attendance has declined, and Catholics eat meat on Friday, practice birth control despite their church's injunction, and are getting divorces freely (if still in not quite the same number as other Americans). The "experts" who specialize in interpreting American Catholicism for non-Catholic readers have rushed to provide explanations. The old Catholic culture, we are told by such Catholic writers as Garry Wills, Wilfred Sheed, Daniel Callahan, John Cogley, and the younger Michael Novak, could not survive the twin modernizing pressures of the Second Vatican Council and the acculturation of the immigrants. Catholics have become better educated, they now think for themselves; the Vatican Council, which started out as an attempt to bring the church into the modern world gently, has ended up destroying most of the old Catholic culture. From the point of view of the non-Catholic reader, the acculturation that he has long expected has finally happened. Education and Americanization are in the process of producing a version of Catholicism which, if not altogether acceptable to the old nativist, at least is not particularly objectionable.

Change is certainly going on in the relationship between the American Catholic collectivity and the institutional church, and it is going on at a rapid rate. But the "experts" are unencumbered by data; indeed, they are quite innocent of it. They have misunderstood the cause and badly described the phenomenon. Corporate Catholicism has managed to get itself into a catastrophe, but it has little to do with the Vatican Council and by no means guarantees that Catholics are "becoming just like everyone else."

Two studies of the National Opinion Research Center (NORC) in 1963 and 1974 enabled us to measure quite precisely the magnitude of change in Catholicism. In 1963, 71 percent of the Catholic respondents reported weekly mass attendance.[1] That proportion has now fallen to 50 percent.[2] Those going to church "practically never" or "not at all" have increased from 6 to 12 percent, and those going to confession "practically never" or "not at all" have increased from 18 to 30 percent. Visits to the church to pray at least once a week have declined from 23 to 15 percent, and daily private prayer has fallen from 72 to 60 percent. The proportion who "never pray," however, remain low at 4 percent; the proportion who pray at least once a week continue to be a quite high 82 percent.

Many of the traditional forms of religious behavior have also declined. The percentage of Catholics who attended a retreat in the last two years has fallen from 7 percent to 4 percent over the decade between the two surveys. The percentage of those who made a day of recollection has fallen from 22 percent to 9 percent, making a mission has fallen from 34 percent to 6 percent, reading a Catholic newspaper or magazine has fallen from 61 to 56 percent, and having a religious conversation with a priest has fallen from 24 to 20 percent.

However, some of the newer forms of religious life that were infrequent a decade ago have attained a surprising popularity. Six percent have attended a charismatic or pentacostal prayer meeting during the last two years, 8 percent an informal liturgy at home, 3 percent a marriage encounter, and 20 percent report having attended a religious discussion group.

The most notable positive change is an increase in the proportion receiving weekly communion—from 13 to 26 percent. Another way of putting this is that less than one-fifth of the weekly mass attenders received communion a decade ago; now more than half of the weekly churchgoers do.

Only 53 percent of the Catholic population thinks that it is "certainly true" that it is a sin for a Catholic to miss weekly mass if he or she could easily attend. The principal reasons for not going to church, however, seem to have little bearing on dissatisfaction directed at the new liturgy. Those who attend mass less than once a week were asked why they do not go to church more frequently, and only 4 percent said they "do not get anything out of mass," while only 7 percent mentioned that they do not like the changes in the mass. (Respondents were able to give as many reasons as they wished for nonattendance. In this analysis we discuss only the most often mentioned first reasons.) The principal reasons for not going to church today seem the same as a decade ago: 10 percent of those who do not go cannot get there because they are too old, too sick, or the church is too far away. Nineteen percent cite laziness or a lack of energy, 14 percent say they have to work on Sundays, and 14 percent say they simply do not want to go.

There has been a substantial decline in acceptance of the legitimacy of ecclesiastical authority. In 1963, 70 percent thought that it was "certainly true" that Jesus handed over the leadership of his church to Peter and the popes; ten years later that proportion has fallen to 42 percent. Only 32 percent think that it is "certainly true" that the pope is infallible when he speaks on matters of faith and morals.

In terms of personal faith, only 38 percent say that they feel "very sure" when they speak to their children about religious beliefs and values. In 1973, 27 percent of the Catholics thought that it was "certainly true" that God would punish the evil for all eternity, a decline of 25 percentage points in the last decade. Thirty-eight percent thought that it was "certainly true" that the devil existed, while 26 percent thought it was "probably true." Still, 86 percent have never thought of leaving the church, 83 percent are married to other Catholics, and 82 percent were married by a priest (down 5 percent since 1963). Despite their own endogamy, the proportion who think it "very important" for young people to marry someone within their own religion has fallen from 56 percent to 27 percent in the last ten years, and the proportion who think it is "not important at all" has tripled to 40 percent.

Loyalty to the church remains, but it is being transformed. A lot of things appear to be not nearly so certain or important as they used to be. But it would be a mistake to think that there is an overt revolt against religious leadership. We asked our respondents whether they approved of the way the pope, the bishops, and their parish priests were handling their jobs. We used the wording that Gallup uses to measure support for the American president and discovered that the local parish priest has a much higher rating than the average that any American president has enjoyed for the last forty years. The pope is one percentage point more popular than John Kennedy was, and slightly ahead of Franklin Roosevelt and Dwight Eisenhower. The bishops are the least popular of the church leaders, with only a little better than three-fifths approving of the way they handle their jobs. Thus the hierarchy ranks beneath Kennedy, Roosevelt, and Eisenhower in popularity, but ahead of Johnson, Nixon, and Truman.

It is clear that religious leadership is much less important than political leadership for most people, and they are much less likely to have strong feelings on the subject. Still, whatever the antipathy may be for certain church policies and whatever decline in confidence there is in certain church doctrines, we do not see in our data much evidence of a strong disaffection from ecclesiastical leadership.

Other researchers have documented changes in birth control practices and attitudes among American Catholics.[3] This change is merely one indicator of a comprehensive shift in Catholic sexual values. Ten years ago, only 29 percent agreed strongly with the notion that husband and wife may have sexual intercourse for pleasure alone. That proportion has now risen to 50 percent. Remarriage after divorce was approved by 52 percent a decade ago; it is now approved by 73 percent. Artificial contraception was approved by 45 percent a decade ago, and now approved by 83 percent. Sexual relationship between an engaged couple was approved by 12 percent in 1963; now it is approved by 43 percent. In 1963, 41 percent thought that "a family should have as many children as possible and God will provide for them"; today only 18 percent would agree with that statement. Eighty percent approve of sex education in Catholic schools. Seventy-two percent thought that abortion should be legal if there was any danger of a handicapped child.

It does not necessarily follow that approval of the positions described above indicates someone is prepared to engage in them himself (or herself). When asked "whether or not you think it should be possible for a pregnant woman to obtain a *legal* abortion," 36 percent said "yes, if the woman is married and doesn't want any more children." When asked what they would do themselves in such a situation, only 8 percent of the women said they would definitely have an abortion, and another 19 percent said they would consider it. (The questionnaire item was, "Now imagine that you are married and you become pregnant, but you and your husband have serious reasons for not wanting to have another child.")

In theory, and to some extent in practice, a substantial proportion of the Catholic population has turned away from what is still the official sexual teaching of the church. This is dramatically pointed out in the decline of the numbers who believe the church has the right to teach what views Catholics should take on birth control. In 1964, 54 percent saw the church as having this right; in 1974, that figure dropped to 32 percent.

The proportion supporting the church's right to teach on racial integration has declined from 49 to 37 percent. Those who support the church's right to teach on immoral books and movies have fallen from 86 to 60 percent. On the other hand, there has been an increase to 51 from 43 percent for those who think the church has the right to teach on the matter of federal aid to education.

The question naturally arises as to whether the tremendous economic and educational progress of Catholics, which was documented in Chapters 3 and 4, explains the dramatic changes in religious practices and attitudes in the last ten years. It would seem almost self-evident that educational and economic acculturation should lead to religious acculturation. However, the educational explanation does not seem to be sustained by the available evidence. First of all, most American Catholics over thirty have not notably improved their personal educational attainment in the last decade. Thus, if the increase in educational attainment of American Catholics were the explanation of the religious changes, then those changes would have been concentrated in the generation under thirty, who are almost entirely responsible for the educational increase among Catholics since 1963. It is certainly true that the younger generation is less religious than its parents, but 80 percent of the change in American Catholic practice has taken place among the generations over thirty. Indeed, educational attainment is not an important predictor of lower levels of religious behavior even among those under thirty.

A more general educational explanation of the change in Catholic behavior would be that the educational advances of the last several decades have produced a substantial segment of the Catholic population sufficiently sophisticated intellectually to grow more weary with the authoritarian rigidities of the church and hence, under the liberating influence of the Second Vatican Council, to revolt against these pressures. However, the evidence shows that there is virtually no relationship between educational attainment and the decline in religious practice. The decline is spread rather evenly throughout all educational levels of the Catholic collectivity. As seductive as the education explanation may be, it simply does not seem to fit the date.

Nor is the decline in Catholic religious behavior in the last ten years the result of a "backlash" against the Second Vatican Council. This notion has been vigorously pushed by some extreme right-wing groups in the church and given encouragement by some Catholic liberals. The laity, it has been suggested by both groups, have been "turned off" by the changes in the church and are protesting with their feet by turning away from the church door. In fact, most Catholics like the new church. More than four-fifths approve of the English liturgy; approximately two-thirds approve of the guitar mass, lay clothes for nuns, and progressive religious education; four-fifths approve of sex education in Catholic schools. In response to the question, "All in all, as far as you are personally concerned, do you think the changes in the church have been for the better, the worse, or don't make much difference one way or another?," 67 percent said the changes have been for the better, 19 percent said worse, and 14 percent said the changes made no difference.

Furthermore, majority support for the changes was found in all ethnic groups and at all educational levels, and was particularly strong among those who had attended Catholic schools for ten years or more (83 percent of this group said the changes were for the better).

Support for the Vatican Council correlated positively with a number of measures of religious devotion. A factor composed of favorable attitudes toward the Second Vatican Council correlates positively with mass attendance, communion reception, confession, daily prayer, accepting the church's right to teach, Catholic activism, and approval of one's son becoming a priest (Table 7.1). Only in matters of sexual and doctrinal orthodoxy is there no relationship between support for the conciliar changes and religious practice, and even here there is simply no relationship rather than a negative one. The Vatican Council did not turn Catholics off religiously, and support for its changes does not lead to lower levels of religious practice despite all the conventional wisdom expressed to the contrary.

TABLE 7.1

*Correlations Between Favorable Attitudes on
Vatican II and Religious Attitudes and Behavior*

	VATICAN II
Mass attendance	.20
Communion reception	.21
Confession	.12
Prayer	.09
Acceptance of church's right to teach	.09
Catholic activity	.20
Sexual orthodoxy	ns*
Doctrinal orthodoxy	ns*
Pleasure at son being a priest	.13

*ns = correlation not statistically significant.

It must be confessed that the NORC research team began its second Catholic study imbued with the conventional wisdom themselves. In the proposal describing the study, we linked the decline of Catholic practice with the Council. However, another major event transpired in Catholic life during the 1960s—publication of the encyclical letter *Humanae Vitae*, renewing the church's prohibition of "artificial" birth control. It therefore seemed appropriate to ask whether it might be the encyclical and the reaction to it that focused intense dissatisfaction with the church's sexual teaching, and thus led to the decline in religious practice.

A "D-systems" model was developed to test this possibility. With-

TABLE 7.2

Changes in Variables Between 1963 and 1974
(percent)

	1963	1974	DIFFERENCE
Mass attendance weekly	71	50	−21
Very pleased with son a priest	66	50	−16
Orthodoxy scale*	45	22	−23
Daily prayer	72	60	−12
Monthly confession	37	17	−20
Active Catholic scale[†]	45	31	−14
Sexual orthodoxy[‡]	42	18	−24

SOURCE: Greeley, McCready, and McCourt, *Catholic Schools in a Declining Church*, p. 125.

*Two or more items on scale composed of "definite proof of God's existence," "evil punished for all eternity," "God cares about how He is worshipped."

[†]Four or more items on scale composed of conversation with priest, frequent communion, above average contribution to the Church, frequent prayer, Catholic TV, Catholic magazines, Catholic books.

[‡]Accepts Church's position on two items (e.g., divorce, birth control, premarital sex).

out attempting an elaborate description of such a model, one can say that it enables the researcher to see how much of the change in one variable over time can be attributed to a change in another variable (Table 7.2).[4] Thus one can "decompose" the differences in behavior at two points in time into the various changes in other variables which seem to account for the change being decomposed. Mass attendance, for example, has declined by 23 percentage points between 1963 and 1964. Twelve percentage points of that change can be linked to the change in sexual orthodoxy in the last decade (Table 7.3).* Seven percentage points of the change can be attributed to the influx of the Vietnam cohort, with their lower levels of mass attendance, into the adult population, and 3 percentage points more can be attributed to the lower scores on sexual orthodoxy of the Cold War cohort. In other words, some 65 percent of the change in mass attendance in the last decade is linked to a change in the sexual orthodoxy of the Catholic population.

There is a strong linkage, then, between the decline in sexual orthodoxy among American Catholics and the declines in church attendance, in happiness over the possibility of a son's being a priest (80 percent of

* When the computer program (named "catfit") used for our models pools coefficients showing relationships in time 1 with the same relationships in time 2, it ignores differences that are not statistically significant. As a result, there is some variation in the percentage point changes in the actual data (presented in Table 7.2) and the model data. The difference of weekly mass attendance, for example, is 21 percentage points in the actual data and 23 percentage points in the modeled data (Table 7.3).

TABLE 7.3

Explanation of Changes in Religious Attitudes and Behavior by Cohort and Sex Attitudes Model
(percent)

	WEEKLY CHURCH	PLEASED WITH SON A PRIEST	DAILY PRAYER	CONFESSION	ACTIVE CATHOLIC
Change due to cohort replacement					
Direct	30	23	30	14	41
Indirect (through sex orthodoxy change)	13	17	20	14	11
Change due to change in sex orthodoxy	52	60	50	45	48
Unaccounted by model	5	0	0	27	0
Total	100	100	100	100	100

SOURCE: Greeley, McCready, and McCourt, *Catholic Schools in a Declining Church*, p. 127.

that change, including indirect transmittance through cohort replacement, being linked to the change in sexual orthodoxy), in daily prayer (70 percent), in monthly confession (59 percent), and in Catholic activism (59 percent). Also, our social change model presented in Table 7.2 is quite successful in accounting for the decline in Catholic religiousness since 1963. In three of the variables—wanting a son to be a priest, daily prayer, and Catholic activism—it accounts for all the changes.

The causal flow could go in either direction, of course. Because one was attending church less frequently, praying less, going to confession less often, contributing less to the church, and less pleased at the thought of one's son becoming a priest, one might be more likely to think there was nothing wrong with birth control or divorce, rather than the other way around. However, it is much more likely that decline in the acceptance of the sexual ethic and decline in happiness over a son's becoming a priest, for example, are linked together to some more basic alienation from Catholicism. But whatever the structure of that alienation, it is clear from Table 7.2 that it is strongly linked to the decline of sexual orthodoxy.

What might this more general alienation be? The most obvious explanation would be that the church's credibility as a teacher, with the right to impose obligations on its members, has been called into question by modern Catholics. Such deterioration of credibility could be either general or specific. It might be linked to a fundamental rejection of the church's right to teach, or it might be linked more specifically to a decline in the credibility of the papacy. If the first possibility turned out to be true, then we would be forced to review the credibility of the

Council model, but if the latter is the case, we will have to turn to the papal encyclical explanation.

To test these alternate possibilities, we used two items that were asked of our respondents in both 1963 and 1974. The first deals with the church's right to teach on matters of racial integration and the other with whether "Jesus directly handed over the leadership of his church to Peter and the popes" (Table 7.4). Racial integration is a controversial subject on which the church has spoken explicitly in the last decade. As we noted earlier, even though Catholics have become more tolerant on racial matters, they have also declined in their willingness to concede the church's right to dictate their attitudes in that area. Such a decline would seem to be a reasonable indicator of deterioration of the church's general credibility as a teacher. On the other hand, acceptance of the pope as successor to Peter and as the Christ-appointed leader of the church seems to be a reasonable indicator of papal credibility.*

TABLE 7.4

Changes in Attitude Toward Church Authority
(percent)

	1963	1974	CHANGE
Church has a right to teach on racial integration (% "yes")	49	37	12
Jesus directly handed over the leadership of his church to Peter and the popes (% "certainly true")	70	42	28

SOURCE: Greeley, McCready, and McCourt, *Catholic Schools in a Declining Church*, p. 127.

While there has been a decline in the willingness of Catholics to accept the church as teacher, this decline does not account for much of the deterioration on other measures of religious behavior and practice (Table 7.5). Indeed, only about 5 percent of the changes in the six variables we are analyzing can be explained by the decline in general teaching credibility, as measured by the church's right to teach on racial matters.

However, when acceptance of the pope as successor to Peter and head of the church is introduced into the change model, a substantial difference occurs (Table 7.6). The percentage of change in dependent variables accounted for by a change in belief in papal leadership is 32 percent for mass (including the indirect transmittance of cohort replacement), 43 percent for happiness with the son becoming a priest, 32 percent for daily prayer, 22 percent confession, and 29 percent for the

---

* An analysis was also attempted using the doctrinal orthodoxy scale as the intervening variable. Since there was little difference between the impact of this scale and the right to teach-racial integration variable on our model, we discuss here only the models using the latter indicator.

TABLE 7.5

Explanation of Changes in Religious Attitudes and Behavior by Model of Cohort, Sex Orthodoxy Change, and Decline in Teaching Authority as Measured by Racial Teaching
(percent)

	MASS	PLEASED WITH SON A PRIEST	DAILY PRAYER	CONFESSION	ACTIVE CATHOLIC	CONTRIBUTION
Cohort replacement						
Direct	29	23	25	14	41	23
Indirect (through sex orthodoxy change)	14	14	20	14	9	6
Sex orthodoxy change	47	57	50	40	46	23
Race teaching decline	5	6	5	5	4	5
Unaccounted by model	5	0	0	27	0	43
Total	100	100	100	100	100	100

SOURCE: Greeley, McCready, and McCourt, *Catholic Schools in a Declining Church*, p. 127.

TABLE 7.6

Explanation of Change in Religious Attitudes and Behavior by Model of Cohort,
Sexual Orthodoxy Change, and Papal Belief Change
(percent)

	MASS	PLEASED WITH SON A PRIEST	DAILY PRAYER	CONFESSION	ACTIVE CATHOLIC
Cohort replacement					
Direct	20	7	30	9	24
Indirect (through sex orthodoxy change)	12	3	5	14	5
Indirect through pope change	5	3	7	4	5
Sex orthodoxy change	32	47	33	41	42
Pope change	27	40	25	18	24
Unaccounted by model	0	0	0	14	0
Total	100	100	100	100	100

SOURCE: Greeley, McCready, and McCourt, *Catholic Schools in a Declining Church*, p. 128.

active Catholic scale. When we look at sexual orthodoxy with the addition of the papal leadership variable we see that the "explanatory power" of sexual orthodoxy shown in Table 7.3 decreases in Table 7.6—20 percentage points for mass attendance, 13 percentage points for happiness at son being a priest, 17 percentage points on daily prayer, 4 percentage points on the confession scale, and 6 percentage points on the active Catholic scale. In other words, change in sexual activities and change in attitudes toward the pope are closely linked in accounting for the decline of Catholic religiousness.

We can therefore draw the following conclusions:

1. The decline in Catholic behavior and practice is linked with the decline in acceptance of papal leadership.
2. The decline in the acceptance of papal leadership is linked to the decline of the acceptance of the church's sexual ethic. For the purposes of the present phase of our arguments, it does not matter which way the causality flows—whether the decline in acceptance of the sexual ethic has led to a decline in papal credibility, or vice versa; nor is it necessary to assert that all the change in the acceptance of papal leadership is explained by the change in acceptance of sexual teaching, or vice versa. All that matters here is that there is a link between the two variables and that the two are linked both independently and jointly to the decline in Catholic belief and practice since 1963.
3. When the papal variable is introduced into the change model, it replaces some of the "explanatory power" of the sexual orthodoxy variable. The decline in Catholic religiousness, in other words, is in

part the result of a joint decline of acceptance in the pope as leader in the church and acceptance of the church's sexual ethic.

We have obviously leaned toward the encyclical explanation as opposed to the Council explanation. We could find no evidence to link the Council to the decline in Catholic belief and practice; we found substantial evidence linking that decline to a rejection of the church's sexual ethic and erosion of the credibility of papal leadership. In the absence of panel data, we cannot say with absolute certainty that people first begin to be less happy about their son being a priest and to pray less frequently and then turn against the Catholic sexual ethic and papal credibility, but the probabilities seem high that the causality flows in the opposite direction: one disagrees with the church's sexual teaching, rejects the authority of the leader who attempts to reassert that teaching, and then becomes alienated from other dimensions of religious belief and practice.

Whatever the causal flow may be, there is nothing in our evidence to suggest that the Council caused a change in Catholic religious practice, and a great deal to suggest that the birth control encyclical caused the decline. Indeed, the evidence in favor of this latter explanation is very strong.

There is another test that we can make. No questions were asked in 1963 about conciliar changes, because none of them had yet been implemented. However, our respondents were asked whether they would support the reform in which the mass would be said in English. We thus had measures in both points in time of attitudes toward the English liturgy, although the wording is necessarily different, since one question was asked before the fact and the other after. We can also separate two of the three components of the sexual orthodoxy scale—attitudes toward divorce and attitudes toward birth control. If these three variables are put into a social change model (leaving out cohorts for the sake of simplicity), we can compare the impact on religious practice and behavior at both points in time of conciliar change for sexual attitudes not linked explicitly with *Humanae Vitae* and for sexual attitudes about which *Humanae Vitae* was written.

Table 7.7 shows this second change model as applied to the variables of mass attendance, support for son being a priest, daily prayer, monthly confession, and Catholic activism. We can see that all of the decline in mass attendance can be accounted for by the social change model II—48 percent attributable to birth control change, 26 percent to change in divorce attitudes, and 26 percent to change in attitudes about papal leadership. We find that birth control accounts for 30 percent of the decline in daily prayer, 38 percent of the decline in monthly

TABLE 7.7

Change in Religious Attitudes and Behavior 1963-1974 as Accounted for by
Social Change Model II
(percent)

	MASS	SUPPORT FOR SON A PRIEST	DAILY PRAYER	MONTHLY CONFESSION	CATHOLIC ACTIVISM
Percent accounted for by English liturgy	0	0	0	0	0
Percent accounted for by change in divorce attitudes	26	19	33	16	29
Percent accounted for by change in birth control attitudes	48	56	30	38	42
Percent accounted for by change in attitudes toward pope	26	25	25	16	29
Percent unaccounted for by the model	0	0	12	30	0
Total change	100	100	100	100	100

SOURCE: Greeley, McCready, and McCourt, *Catholic Schools in a Declining Church*, p. 135.

confession, and 42 percent of the decline in Catholic activism, while divorce change accounts for a third of the decline in prayer, 16 percent of the decline in confession, and 29 percent of the decline in activism. It should be noted that the nature of the model is such that these relationships are *net*; that is, they represent the influence of a change in birth control, taking into account any related or overlapping change in divorce attitudes.

The change in birth control thinking, then, is clearly the most important factor at work in the decline of Catholic devotion and practice during the last decade, with related declines in divorce and respect for the papacy combining with birth control to account for *all* the deterioration in mass attendance, support for a priestly vocation in one's family, and Catholic activism. In the remaining two cases (prayer and confession), the model accounts for most of the change. Only 12 percent of the change in daily prayer and 30 percent of the change in monthly confession cannot be attributed to the factors at work in the model. None of the change is attributable to the English liturgy—the only available measure of attitude toward the Vatican Council at both points in time.

There was, in other words, a dynamic built into the events of the last decade which would have led to an increase in Catholic religiousness if it had not been for the deterioration in the sexual ethic and support for the papacy. Given the positive response to the changes instituted by the Council, it is not unreasonable to assume that the Council is at least in part connected with such a dynamic. It is very likely that if it had not been for the positive dynamic introduced by the Council, the deterioration analyzed in this chapter would have been even worse.

In order to separate the positive dynamic, associated with the frequency of communion reception and linked to the Council, from the negative dynamic, associated with the decline in sexual orthodoxy and support for the pope, we developed social change model III. The three "internal" variables—the advent of the cohort under thirty, the decline in acceptance of papal leadership, and the decline in sexual orthodoxy—represent the negative dynamic. If the direct path from time is positive, it represents a positive dynamic (Figure 7.1). When the increased weekly reception of communion is put into the model (bracketed), it should lead to a decline in both the direct path from time and in the indirect path through the three internal variables, if increased reception of communion represents a positive dynamic. The net change must be whatever the actual decline in the variable being measured has been over time.

If we apply this test to the active Catholic scale (Table 7.8), we see that without considering the influence of Holy Communion, the changes "internal" to the model would have led to a decline of 21 percentage points between 1963 and 1974 (from 45 to 24 percent). (This is one and a half times greater than the actual decline, which is why the "total" decline in Table 7.8 equals 150 percent.) However, the positive force attributed to the change over time attenuates the decline, so that instead of a 21 percentage drop we see only a 14 percentage decline in the active Catholic scale, a decline which is linked to the encyclical *Humanae Vitae* and was canceled out by a positive force also at work during this time period. The extent to which this positive force is linked with the increased reception of communion and thus to the effects of the Council can be judged by what happens to the positive path from time when communion is introduced into the model in Figure 7.1. If the positive path declines, it means that the increase in communion reception accounts for proportionately more of the positive force mentioned above.

In fact, the positive path does decline; it vanishes at zero when communion reception is brought into the model (Table 7.8, part ii). In other words, the increase in communion reception accounts for all of the positive dynamic at work in social change model III. Therefore, if the Vatican Council had been the sole force at work from 1963 to 1974, the proportion above the mean on Catholic activism would have risen seven points (from 45 to 52 percent). If, on the other hand, the encyclical *Humanae Vitae* had been the sole force, that same proportion would have declined twenty-one points (from 45 to 24 percent). What actually happened was that the two forces operated simultaneously. The larger negative force of the encyclical masked the smaller positive force of

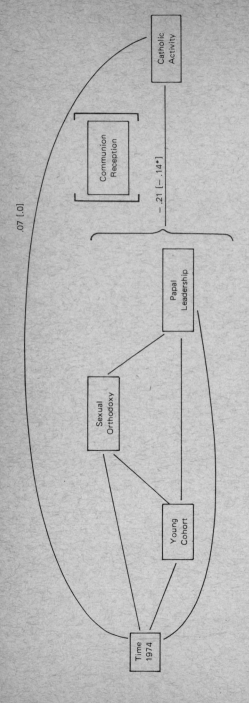

FIGURE 7.1

Social change model III with the impact of frequent reception of communion.

SOURCE: Greeley, McCready, and McCourt, *Catholic Schools in a Declining Church,* p. 138.

* The statistic in brackets indicates relationship when weekly reception of Holy Communion is placed in the model.

TABLE 7.8

Active Catholic Scale and Social Change Model III
(total decline = 14%)

i. WITH INCREASE IN COMMUNION RECEPTION LEFT OUT		
Explained by change in attitude toward pope	— .028	— 20%
Explained by change in sexual attitudes	.144	−102%*
Explained by new cohort	.038	— 27%
Total decline attributed to changes in sex and pope attitudes and cohort change	−21.0	−150%*
Explained by change over time	.700	50%
Actual total change	— .140	100%

ii. WITH INCREASE IN COMMUNION RECEPTION PUT IN		
Total explained by sex, pope, and generation	— .140	100%
Direct path from time	.000	000%
Total change	— .140	100%

SOURCE: Greeley, McCready, and McCourt, *Catholic Schools in a Declining Church*, p. 137.

*The "internal" percentages of part i add to more than 100 because the decline would have been greater if it had not been for the increase in activism over time.

the Council, but the Council had the effect of attenuating the larger negative influence of the encyclical by about one-third.

Left to itself, the Council would have led to an increase of about one-sixth in Catholic religious practice. Left to itself, the encyclical would have led to a decline of almost one-half. The net result is a decline of almost one-third as far as the active Catholic scale is concerned. Far from causing the problems of the contemporary American Catholic church, the Council prevented them from becoming worse.

But did not the Council prepare for a hostile reception of the encyclical? If there had been no Vatican Council, would not the Catholic population have been much more likely to accept the papal decision made by *Humanae Vitae*? At first this seems like a reasonable assumption. However, it may not take into account how aware the typical American Catholic was of the "new spirit" in the church generated by the Council. It seems safe to assume that the Council was not of central concern in the daily lives of most Catholics. It was an interesting and colorful event, no doubt, but they were not listening very closely to the arguments that went on about the nature of authority in the church. It may be that their expectations on the subject of birth control, an issue that was of central concern to many, were raised somewhat by the conciliar atmosphere, only to be decisively shattered by the encyclical. Yet it seems likely that, regardless of the Council, the opening up and then the closing down of the birth control question would produce much the same impact that it did. Furthermore, research done by Westoff and his colleagues at Princeton has shown that birth control practice among Catholics was changing long before the Council, and would have most likely continued no matter what went on at any meeting of bishops in Rome.[5] Quite simply, the invention of the birth control pill called for a formal decision. That decision was made, and the Council was irrelevant to the issue, save for generating more publicity about decision making in the church.

In 1965, 77 percent of Catholic women under forty-five were practicing some form of conception control, 28 percent of them using the church-approved rhythm method. Five years later, after the encyclical, there was only a 4 percentage point increase in conception control (up to 81 percent), but a decline of half in those using the rhythm method (to 14 percent). The proportion using the pill went up from 12 percent to 28 percent; the proportion using other methods besides the pill and rhythm remained unchanged at 37 percent for the five-year period. All of the increased nonconformity involved use of the pill, and much of it resulted from the replacement of rhythm by the pill. It would appear that a large number of Catholic women made up their minds in the late 1960s that the pill was more effective than rhythm and no less immoral. Indeed, the women who were using the pill in the late 1960s were *more* likely to receive Holy Communion at least once a month than those who practiced rhythm or no method of birth control. Twenty-six percent of the former group received communion once a month, as opposed to 23 percent of the latter group, and among those under thirty, 37 percent of the pill users were receiving monthly communion, as opposed to 18 percent of the rhythm users and 15 percent of the no birth control group.

An additional piece of evidence of a change in attitude in the late 1960s can be found in the dramatic increase in monthly reception of communion by pill users. Only 11 percent received monthly communion in 1965, while the figure was 25 percent five years later. For those under thirty the proportion of monthly communicants among pill users increased from 20 percent to 37 percent.[6]

At just about the time *Humanae Vitae* was being drafted (desperate last-minute efforts were being made to stop it, according to Vatican rumors at the time), a substantial segment of American Catholic women were making decisions contrary to the pope's: The pill was not sinful, and its use was not an obstacle to reception of the sacraments. As Westoff and Bumpass note, "It seems clear that the papal encyclical has not retarded the increasing defection of Catholic women from this teaching."[7] Presumably they would have made such a decision with or without *Humanae Vitae*. One can only speculate whether it was counterproductive in its explicit intent and actually led to an increase in the use of the pill by women who were angry and disappointed by the encyclical.

In any case, development of the pill created a new moral situation which the church would have had to deal with whether or not there had been a Vatican Council. Between 1960 and 1965 (before the end of the Vatican Council), unapproved forms of birth control had increased among American Catholics from 38 percent to 51 percent, with almost all of the change during those years being accounted for by the invention of the pill. The change in the second half of the decade was, as we have noted, almost entirely the result of a switch from rhythm to pill. Until 1968 this change had little impact on religious practice, as we shall see shortly. It was only after the encyclical that weekly church attendance began to drop precipitously. The Vatican Council did not produce the birth control pill, did not lead Catholic women to use it before 1965, and can hardly be said to have caused more to use it after 1965. Perhaps all the Council did was to give women more confidence about receiving Holy Communion even when they were using the pill. Thus the Council did not lead to a decline in religious practice, but allowed women to continue to accept the sacraments who otherwise would not. The Council and the pill, then, are relatively unrelated phenomena, save for the fact that the Council may have mitigated the negative effect that the birth control decision apparently has had on religious practice.

The partial alienation of Catholics from the institutional church seems to be almost entirely the result of the unresponsiveness of the church's sexual teaching, particularly as demonstrated by the encyclical letter *Humanae Vitae*. Evidence to confirm this is available from other data sets. Church attendance did not decline between the Council and the issuance of the encyclical letter, but has fallen sharply ever since. The apostasy rate among American Catholics was 7 percent in 1953; it had not changed by 1967, the year before the encyclical. However, by the early 1970s the apostasy rate had doubled to 14 percent, even though there was no change in the Protestant rate (9 percent) during the same years. In 1967, 48 percent of the American Catholic population thought that the church was losing influence in American society. By 1970 that proportion had increased 27 percentage points, compared to a 15 percentage increase for Protestants. In the years between the Council and the encyclical, fewer than 1 percent of the priests in the country had resigned. In the two years immediately after it, almost 5 percent of the priests resigned. Dissatisfaction with the church's sexual teaching undoubtedly was widespread before the encyclical, but when all hope was turned off for reconsideration of the teaching, the encyclical became the catalyst for the dramatic religious change that has occurred in the years since.

The ten years after the end of the Second Vatican Council have naturally been a turbulent time for the church. Powerful forces of both growth and decline have been set in motion. For the moment, decline seems to be the more powerful. If one projects them into the immediate future—ten or fifteen years—what can one expect will happen to the church as an institution?

First of all, it will have much less income relative to the past. According to the calculations of my colleague William McCready, the income of the church in 1974 was $1.7 billion less than a decade previous. If Catholics had contributed the same share of their personal incomes in 1974 that they did in 1963, they would have contributed more than $5.5 billion. In fact, however, their contribution was $3.8 billion. About three-fourths of the loss seems to be the result of negative reaction on the part of American Catholics to the birth control encyclical *Humanae Vitae*.*

NORC estimates of Catholic contributions are based on questions asked of Catholic families in 1963 and 1974 about their annual contribution to the church and multiplied by one-quarter of the families and unrelated individuals in the country. (Virtually all data indicate that Catholics are 25 percent of the American population). In 1963 the average Catholic family gave $164 to the church. This produced an estimated church income of $2.6 billion. By 1974 the average contribution had increasd to $180 a year. Taking into account the increased size of the Catholic population in the last ten years, this contribution resulted in an estimated income of $3.8 billion—an apparent increase of $1.2 billion.

However, as every American knows, the dollar is not worth now what it was in 1963. Inflation has substantially reduced its purchasing power, so what looks like an increase actually becomes a decrease. To have kept pace with inflation, Catholics would have had to increase their annual contributions to $262 a year, thereby producing approximately $4.2 billion a year in income. Hence the failure to keep pace with inflation has in fact produced not an increase of $1.2 billion but a decrease of about $400 million of income—a decline of about 10 percent in income in inflation-free dollars.

But the real picture may be even worse. In 1963 Catholics gave about 2.25 percent of their family income to the church. Despite inflation, the real income of Catholics (in inflation-free dollars)—like that of all Americans—increased substantially since 1963. Catholic contributions not only failed to keep pace with inflation, they also failed to

---

* Information on Catholic finances is difficult to acquire because there is no central accounting system. The decentralized financial control of the church has created a situation in which even the hierarchy has only a guess, at best, as to what the collective contributions of American Catholics are.

keep pace with the rising Catholic standard of living. Hence in 1974 Catholics were giving only a little more than 1.5 percent of their family income to the church—a decline of 31 percent. If Catholics were contributing at their 1963 rate in 1974, the annual income of the church would have been more than $5.5 billion. Hence, despite the apparent gains, the Catholic church has in fact suffered a decline of almost one-third in its annual income during the last decade.

Many church leaders insist that "contributions are up," and deny that there is any financial loss. But this argument ignores both the decline in the value of the dollar and the increase in Catholic real income. What are the reasons for the deterioration of contributions? Some conservative Catholics claim that it represents a revolt against the innovations brought into the church by the Vatican Council. But NORC's computer models show that only about 6 percent of the decline seems to be the result of a conservative backlash. Nor does the decline in mass attendance seem to be the principal explanation; only 7 percent of the decrease in income is accounted for by the lower proportion of Catholics who go to church every Sunday.

As for virtually every other serious decline in Catholic activity during the last ten years, the explanation seems to be a strongly negative reaction to the encyclical letter on birth control, *Humanae Vitae*. Seventy-three percent of the decline in contributions can be accounted for by changing attitudes toward papal authority and sexual morality. In dollars-and-cents terms, the birth control encyclical seems to have cost the church about a billion and a quarter dollars.

The average cost of Catholic schools for parents of students in 1974 was $343 (or 3.32 percent of their income). Since some 2.2 million families have children in parochial schools, the total money expended on direct costs for Catholic education in 1974 was approximately $800 million (though of course Catholic schools are not paid for completely by tuition but are subsidized, to an extent that no one has been able to calculate, out of general parish revenues). If one combines contributions to the church with parochial school costs, the total income of the Catholic church in the United States in 1974 was approximately $4.6 billion.

There is substantial potential support for Catholic schools that has not yet been tapped. Eighty percent of the NORC respondents said they would increase their annual contribution to the church if it was necessary to keep their parochial school open. Fifty percent said they would give more than $50, and 25 percent said they would give more than $100. The average contribution to save the parochial school would be $105 per year, or about $1.8 billion from the entire Catholic population. In other words, there is apparently enough money available for support of Catholic schools in a crisis situation to make up for the financial losses the church has incurred through declining Sunday contributions.

If all the Catholics who are willing to make contributions to save their parish school would in fact do so, the income of the church would be about $5.6 billion (in addition to money used in direct payments to the schools from those who have children in them).

Such increased support for parochial schools does not seem unrealistic, since it would merely mean that Catholics would return to the level they had reached in 1963 of proportionate income given to the church. Obviously there would need to be major changes in the way the parochial schools are administered and financed before such increased contributions might reasonably be expected.

The use of survey data to estimate church income is no substitute, of course, for systematic national accounting and accountability. But until such systems are developed, the survey data at least give approximate information about the state of Catholic school contributions and parochial school finances. One must conclude that the crisis in parochial schools is much worse than it need be.

Church attendance, support for a priestly vocation, private devotion, financial contributions—all these important indicators of Catholic religious orthodoxy have declined dramatically in the last decade because of the influence of the revolt against the church's sexual teaching. Yet the overwhelming majority of those who were born Catholics are still Catholic. Three-fifths of them still pray every day, reception of Holy Communion has doubled in the last ten years, and, as we shall see in subsequent chapters, Catholics still support vigorously the parochial schools and still like (if not respect) their clergy.

But what will happen in the years to come if the trends documented in the last decade should continue? My colleague Garth Taylor has constructed a mathematical model which enabled us to simulate developments in Catholicism through the next decade and a half. These estimates of the future are not guesswork or idle speculation. They are based on computer simulations which take into account the demographic changes in the population as a new generation of young people become adults, and which project observed trends into the future. These projections are not predictions; they are rather "scenarios" of what will happen if certain trends continue. The researcher who works with simulator models normally develops a number of scenarios, within one of which the future is likely to occur unless some outside force intervenes —in this case, a general religious revival in the country, another ecumenical council, repeal of the birth control encyclical, or a startling new papal administration, for example. Scenarios are trend lines into the future which may become self-defeating prophecies if those who have the power to change the trends take their possibility seriously.

If the next fifteen years are as bad as the last ten for the American Catholic church, it will be hardly recognizable as the church which emerged at the end of the Vatican Council, much less the church that existed before the Council. Only one-third of its members will be going to mass every week. A mere 29 percent will find themselves very pleased at the prospect of their son being a priest, and only 12 percent will be giving more than $180 a year (in inflation-free dollars) to the church. A tiny 1 percent will be active in church affairs. The institutional church will be gravely ill, indeed almost moribund. Half the Catholics in the country will continue to pray every day, religion as private devotion will continue to be vigorous, but religion as church activity will be approaching the vanishing point.

The scenario described above is the "worst case" simulation. It assumes that the decline in acceptance of sexual teachings and papal authority, which seems to have resulted from the birth control encyclical, will continue at the same pace for the next fifteen years as it has for the last ten. It also assumes that the new generation of young people who will become adults in the next fifteen years (those who are now from two to seventeen years old) will be as much less religious than their parents as parents today are less religious than theirs. Under the circumstances of the "worst case," weekly church attendance will fall 40 percentage points between 1963 and 1989, support for a priestly vocation for one's son will drop 33 percentage points, daily prayer will fall 21 percentage points, and contributions over $180 will decline 38 percentage points. The quarter century between 1963 and 1989 will record one of the worst church disasters in human history—almost entirely because of opposition to the birth control encyclical. The "worst case" is by no means an improbable outcome.

A less gloomy scenario cuts the rate of decline in acceptance of papal authority and birth control teaching in half, and sees no difference in religious behavior between today's younger generation and tomorrow's. But even then weekly church attendance will decline to 40 percent (from 72 percent in 1963), support for son a priest to 38 percent (from 62 percent), daily prayer to 53 percent (from 72 percent), Catholic activism to 13 percent (from 50 percent), and contributions of $180 or more to 16 percent (from 50 percent in 1963). Even a notably reduced rate of decline will still constitute a severe trauma for American Catholicism.

It may be that the worst of all the ill effects of the last decade are over, and that the decline of Catholicism will "bottom out" in the next fifteen years. But such a bottoming out will not mean a turnaround. On the contrary, because young people are now much less devout than their elders and because they will constitute a greater proportion of the population in years to come, some erosion will continue. In the bottoming out scenario, weekly mass attendance in 1989 will have stabilized at 48 percent, the proportion of being "very pleased" at having a son choose the priesthood will also be 48 percent, and 58 percent will be praying every day. Twenty-four percent will be activists, and 30 percent will be contributing more than $180 (inflation-free) a year to the church. The church will not have recaptured any of its old vitality, but neither will it be threatened with collapse.

Are there no optimistic scenarios? There were two forces at work during the past decade within American Catholicism—a negative force associated with the birth control encyclical, measured by declining support for papal authority and the official sexual morality, and a positive force associated with the Vatican Council, measured by increasing reception of Holy Communion every week. What if the negative force bottoms out and the positive force continues to grow, while half of the youthful drift away from the church is turned around?

Under those very optimistic circumstances, the revival in American Catholicism would not be spectacular. Mass attendance would rise from its present 50 percent to 56 percent, daily prayer from 60 percent to 67 percent, activists from 28 to 34 percent, and contributors of more than $180 from 38 to 40 percent. Thus, even in the most hopeful set of circumstances it seems reasonable to imagine (and this would include an extraordinary increase in the reception of Holy Communion) that the church would be able to reclaim only a modest amount of the losses it suffered between 1963 and 1974 (most of them occurring in the shorter span between 1968 and 1974).

The dynamics at work in American Catholicism at the present time are such that it is relatively easy to imagine the "worst case" eventuality and relatively difficult to imagine the quite modest "best case" eventuality. The birth control encyclical not only canceled out the effects of the Vatican Council, it also set into motion forces which have caused grave losses to the Catholic church and which will be very difficult to reverse in the next fifteen years.

Such scenarios are not prophecies, but they are certainly warning signals. Something rather like our more pessimistic scenario will occur in American Catholicism during the next fifteen years unless some outside force intervenes to break up the dynamics in our models—dynamics which merely reflect what has happened during the last ten years. It is possible that forces external not only to the model but external to the church will intervene. There might be one of the periodic great religious revivals which seem to sweep across American society when they are least expected. Surely Catholicism would not be immune to such a revival.

But it would be naive to place all of one's organizational hopes in such a revival. The leadership of American Catholicism will have to create their own external intervention if they have any concern at all about the future of the church. (One bishop remarked when told of our findings, "What do I care about 1989? I'll be dead by then.") Our scenarios cannot take into account the possibility of such intervention. But it is a melancholy commentary on the decline of American Catholicism that they probably don't have to.

The Catholic church in the United States fifteen years ago seemed strong and vigorous. Seminaries and novitiates were filled, vocations to the priesthood and the religious life were increasing, as was attendance at Catholic schools. Organizational activity was thriving. The Vatican Council appeared to be an extremely successful effort at modernization; the ecumenical age had begun in a burst of enthusiasm. A decade and a half later, the church is in an organizational shambles. Its apostasy rate has doubled, church attendance has declined, financial contributions have gone down by a third, thousands of priests and nuns have left the ministry, and the prospects for the future are hardly encouraging. Virtually all of this decline, it would seem, is the result of a single problem—birth control.

The international leadership of the church succeeded in persuading itself that concern about family planning was the result of selfishness, a "contraceptive mentality," a lack of respect for life, and a deterioration of the dignity of marriage. It was unable to grasp that the rapid progress in dealing with infant mortality had created a world population explosion and a situation in modernized countries in which a woman could easily have seven or eight children before she was thirty. An international commission of experts had been assembled and had developed a social and ethical theory which would have justified modification of the birth control teaching; but backstairs Vatican politicians persuaded the pope to overturn the decision of his own commission and issue the encyclical *Humanae Vitae*. Desperate last-minute attempts to tone down the encyclical failed, and the Catholic world awoke on a midsummer day in 1968 to find that the Vatican leadership had never heard what it was saying. Some bishops responded enthusiastically, but most bishops in the United States were perfunctory in their response and signaled to their clergy that it was all right to continue to leave birth control decisions to the Catholic married couples themselves. This was a necessary decision under the circumstances, but it could scarcely help the credibility of the bishops.

The encyclical was ignored in practice by both the clergy and the laity in the United States. Almost four-fifths of both groups agreed that birth control was not sinful. Little attempt was made by the hierarchy to enforce the encyclical, but still it became a test of loyalty among church leaders. Anyone who expected promotion into the hierarchy or upward within it knew that as long as Paul VI was pope, he had to remain silent. The American bishops knew that they would not be able to enforce the encyclical (although they were probably surprised at how overwhelming its rejection was), yet there is little evidence that they communicated this knowledge to higher levels of ecclesiastical authority. The structure of ecclesiastical authority of the church was such that even those American bishops who personally felt that *Humanae Vitae* was a mistake have been able to do nothing but bide their time and hope that the next pope will take steps to mitigate the disaster.

The American Catholic church as an institution has not been completely undone. It has suffered a severe trauma brought on not by "enlightenment" or secularization or acculturation or even by a revolt against authoritarianism. The disaster for American Catholicism was the result of a single decision made because of the decrepit and archaic institutional structure of the church, a structure in which effective upward communication practically does not exist.

Other American denominations have survived for many years with a much lower level of religious involvement from their members than American Catholicism seems likely to have even in our "worst case" scenario. Other collectivities have survived with much weaker institutional structures than Catholicism is likely to have in the years ahead. Neither as a denomination nor as a collectivity will American Catholicism vanish; but the denomination has suffered a grave setback, and the outcome for the collectivity remains to be seen. As I will suggest in the conclusion of this book, one possible result will be the emergence of many different styles of affiliation with the Catholic collectivity. The boundaries of the church will be wider and more permeable, and Catholics will affiliate with it in the manner of their choosing, and not of the institution's.

There have been more numbers and more complex analysis in this chapter than in any of the others. Unfortunately, any attempt to do justice to the rich possibilities of the new art of social change analysis cannot escape such technicalities. But the result of picking one's way through such a tangle of complexities is a much more sophisticated version of what has happened between two points in time than any simple and undocumented speculative explanation might offer. There were two major dynamisms at work in the American Catholic collectivity during the late 1960s and early 1970s, one flowing from the momentous and extraordinary Second Vatican Council and the other from the negative response to Pope Paul VI's birth control encyclical. The former dynamic was positive and set to work growth-producing reactions; the latter was negative and more than canceled out the impact of the former. But that does not mean that the positive dynamism is not operating; it continues to operate, but its effect is masked by the negative. Is the Catholic church as an institution growing or declining? It is doing both, but the decline is much more obvious.

REFERENCES

1. The 1963 study is reported in depth in Andrew M. Greeley and Peter H. Rossi, *The Education of Catholic Americans* (Chicago: Aldine, 1966).

2. The 1974 study is reported in depth in Andrew M. Greeley, William C. McCready, and Kathleen McCourt, *Catholic Schools in a Declining Church* (Kansas City: Sheed & Ward, 1976).

3. Charles F. Westoff and Larry Bumpass, "Revolution in Birth Control Practices of United States Roman Catholics," *Science* 179 (January 1973): 41–44.

4. For further details of this method of social change analysis, see Greeley, McCready, and McCourt, *Catholic Schools in a Declining Church*, chap. 5.

5. Westoff and Bumpass, "Revolution in Birth Control," pp. 41–44.

6. Ibid., pp. 179–180.

7. Ibid., p. 41.

CHAPTER
8

The Catholic Clergy

IN THE AUTUMN of 1975, for reasons known only to the people who make decisions in the high towers of Manhattan about what the American public will have to like on television this year, a number of programs about explicitly Catholic families began to appear on prime viewing time. For a few months (the shows, for not quite so unknown reasons, were canceled before the season was even half over), we were privileged to see the Lassiters, a historically improbable wealthy family on Beacon Hill; the Montefuscos, an Italian family that gestured a lot and told Jewish jokes; and the McShanes, a father-detective, daughter-lawyer team who handled "relevant" legal problems. Each of the families had a priest attached. "Father Tom" was a Lassiter missionary, Father Montefusco gestured like the other family members but did not have nearly as many laugh-lines, and the McShane Jesuit solved legal problems and shot an effective game of pool. The Bing Crosby, Barry Fitzgerald, Spencer Tracy, Pat O'Brien type of clergyman was alive and well, and it was the 1940s, not the 1970s.

Watching the serene, confident clerics, and interrupted only by the priest who was advertising Datsuns, one could almost forget that Daniel and Philip Berrigan allegedly plotted the kidnapping of Henry Kissinger, and that Philip Berrigan had lived for a number of years in a curious common-law marriage with a nun. One could forget James Cavenaugh's article in the *Saturday Evening Post*, "I'm a Catholic Priest and I Want to Be Married," which was followed by his "A Catholic Priest Looks at the Outdated Church," which was followed by his marriage and divorce. One could forget that thousands of priests had resigned from the priesthood, and that many more seemed to be in a severe identity crisis as they desperately searched for "relevance" in a rapidly changing church where the role of a clergyman was no longer clear and the model for appropriate clerical behavior had become uncertain. If Catholic priests were not as radical as the Fathers Berrigan or as troubled as Father Cavenaugh, were they as poised and self-confident as Fathers McShane and Montefusco?

The available empirical evidence, collected in 1970 for a study done at the National Opinion Research Center (NORC) for the national hierarchy (and subsequently ignored if not disowned by its sponsor), suggested that the vast majority of American priests were somewhere in between.[1] By 1970 some 5 percent of those who were priests in 1965 had resigned from the priesthood, and 3 percent of the NORC national sample said they would certainly or probably leave, while another 8 percent were uncertain about their future. But almost 60 percent of the priests in the country said they would definitely not leave the priesthood. A little fewer than half the priests in the country acknowledged that they had rethought their status because of a friend's leaving. The principal reasons given for staying were a sense of vocation (mentioned by about two-thirds), happiness in priestly work (mentioned by about two-fifths), and giving witness to Christ (mentioned by about one-third). A substantial majority of the clergy (about three-fifths) thought that celibacy should be optional (65 percent of the laity thought it should be optional, and 80 percent of the laity could accept married clergy). About three-fifths of the priests expected a change in the law of celibacy, most thinking it would come within ten years. However, fewer than one-fifth said they themselves would marry. Among the reasons listed for the value of celibacy, 78 percent said that it helped them to do their work better, and more than one-half mentioned personal growth and development, development of love for God, and relating more fully to people. Some 4 percent of the clergy went out with women several times a week, 10 percent several times a month, and 34 percent several times a year. Two-fifths of those who reported socializing with women said they did so as members of a group and not on dates. Eighty-two percent of the clergy said that they either never went out with women or did so only several times a year and in a group.

Their celibate status, however, did not mean that priests had lower levels of life satisfaction than other men of the same age and educational achievement. On the contrary, in each age group (twenty-six to thirty-five, thirty-six to forty-five, forty-six to fifty-five, and over fifty-five) the score on a psychological well-being scale developed by Norman Bradburn[2] showed priests higher in psychological well-being than married males of the same age. They were substantially higher than unmarried males of the same age (who have the lowest psychological well-being scores of all categories). The Bradburn Psychological Well-Being Scale is composed of both "positive affect" and "negative affect," the former indicating positive feelings and the latter negative feelings. The final score on the scale consists of the balance of positive over negative affect (or the reverse). Interestingly, there was little difference between priests and their age peers in positive affect. Most of the differences in psychological well-being were in negative affect; priests and married men seem to have about the same amount of positive feeling toward their lives, but married men of the same age and educational background have substantially more negative feelings.

A personality test (the Personality Inventory [POI])[3] that purports to measure "self-actualization" was administered to a subsample of priests in the NORC study (more than 900 men). Priests scored generally between the self-actualized and non-self-actualized norms of the test. With age held constant, priests had higher scores on self-actualization than college students, Peace Corps volunteers, supervisors in an electronics company, and a sample of New Zealand Baptist ministers. They also had virtually the same scores as a group of divinity school students at the Protestant Colgate-Rochester Divinity School.

Most of the respondents on whom the POI "norm" is based are under thirty-five, and Catholic priests under thirty-five are virtually on the norm with their scores on the test. In particular, a score on a subscale called "capacity for intimate contact" (an indication that the respondent has "warm interpersonal relationships") is well within the normal range. In other words, insofar as the personality test measures abnormality, celibate commitment, closed training, and the peculiar life style of the clergy has not produced abnormal personalities. Statistically, at any rate, there is very little difference between the psychological profiles of American Catholic priests and other American males of comparable education and age.

Only on a measure used to gauge work satisfaction do priests score generally lower than other professional American males. Their work satisfaction is generally about the same as middle managers in industry —with the single exception of associate pastors, whose work satisfaction is about the same as unskilled workers in industry. Nonetheless, four-fifths of the priests in the country report that they would enter the priesthood again, and more than three-quarters compare themselves favorably with other professionals in the depths of knowledge and skill required for their work, in the responsibility they have for an undertaking, and in their commitment to serving the needs of people. The proportion rating themselves favorably on the criterion of autonomy in decision making is only 58 percent, however.

In summary, both by standardized tests and by their own evaluation, there is little evidence that the majority of priests appear deficient in comparison with other males the same age and educational background in American society. They score higher on measures of psychological well-being ("happiness"), about the same on scores of emotional maturity, and about the same on scores of work satisfaction. Their jobs are not as satisfying as those of other professionals, but still priests compare their working conditions favorably with what they perceive to be the working conditions of other professionals. The majority of priests do not intend to leave the priesthood and are happy with the decision they made to become priests. All may not be well within the Catholic priesthood, but it does not seem to be the disaster area that is often portrayed in newspaper and journal accounts.

Why, then, do some men leave the ministry? There is no simple answer because there are many different reasons. The NORC priesthood report developed a complex mathematical model which explained about half the variance in future plans for affiliation with the priesthood. The desire to marry certainly was the precipitating factor in many decisions to leave, but in careful analysis of those who were preparing to leave and those who had already left, it became clear that, with some exceptions, the majority of those who left the priesthood were unhappy with the work they had to do in the ministry. They liked the kind of things most priests do not get an opportunity to do (social and political organizing, for example) and did not like the things that occupy a considerable amount of the time for most priests (visiting the sick, preaching, instructing, working with grammar school children). The majority of those who left the priesthood would not return even if they could do so and still remain with their wives, and only 10 percent would return to the kind of active ministry they had left behind if they could do so as married men.

Personal and psychological well-being and, for married priests, marital satisfactions are very high in the early years after leaving the priesthood, but they diminish sharply in subsequent years. There is no evidence that the low levels of marital adjustment among married clergy are any different from those which obtain among other men who married at the same age as did the resigned priests. Some 84 percent of their wives were Catholic, and 43 percent were members of a religious community. About two-fifths of the resigned priests still consider themselves to be organizational Catholics, another two-fifths think of themselves as Catholics beyond the structure of the organization, and a remaining fifth do not define themselves as Catholics or, in most cases, even as Christians. These proportions are not affected by length of time since resignation—which suggests that for about one-fifth of those who left the priesthood, perhaps, departure was linked to disillusionment with and subsequent rejection of Christianity.

Despite the oft-expressed opinion that it was the "best" who were leaving the priesthood, there is no evidence that the resignees were better educated or scored any higher on measures of emotional maturity than those who remained. In summary, those who leave the priesthood seem to be those who do not like being priests. And their reasons include personal loneliness, frustrations on the job, a desire to marry, a loss of personal faith, or some combination of these and probably other causes we did not tap. There was also some evidence that those who left the priesthood had less happy childhoods and were less likely to have had intense religious or ecstatic experiences than those who stayed.

Despite their relative satisfaction, those who remain in the priesthood vigorously endorse a wide variety of changes in the church. More than three-quarters support a concept of some married priests working in a variety of ministries, an appeals court for the church, election of the pope by international assembly of bishops, and election of bishops by the priests of the diocese.

Sixty percent of the priests do not think that all artificial contraception is wrong, and only 29 percent are certain that it is wrong—a decrease of 11 percentage points from the position held prior to the encyclical *Humanae Vitae* (further evidence of the counterproductive effects of that encyclical). Only one-quarter of the priests in the country in 1970 were willing to refuse absolution to those who would not promise to stop artificial contraception. Forty-two percent thought that the encyclical letter *Humanae Vitae* was a misuse or abuse of authority, and 18 percent more thought it was an inappropriate use of authority. Thus only 36 percent were willing to contend that the encyclical was a competent and appropriate use of authority. One-third of the priests were convinced that divorce was absolutely forbidden by divine law and could not be permitted by the church, though 80 percent of the clergy thought that premarital sex was always wrong. On divorce and birth control, at any rate, there is very little to choose from in the attitudes of the majority of the Catholic clergy and the Catholic laity; both are in substantial disagreement with the official teaching of the institution.

There were actually more American priests in this country in 1975 (about 58,000) than there were in 1965, in part a function of large ordination classes in recent years due to the large age cohort in its late twenties, and in part a function of including missionary priests working outside the country in the total number of priests. On the other hand, there has been an enormous decline in the number of seminarians—50,000 in 1965, fewer than 18,000 at the present time. Clergy admit that they are themselves much less likely to encourage young men to enter the priesthood; they do not discourage such vocations, but simply leave a good deal of the choice to the individual, failing to offer the positive reinforcement they were likely to offer in the past. Laity, as we have seen, are substantially less likely to say that they would be "very pleased" if their son chose to be a priest. Indeed, the priesthood as a profession for one's son is about as popular as a son's being a business executive or a stockbroker; it is substantially less popular than a college professorship.*

The turbulence of the Vietnam era and the resulting disaffiliation from all institutions by that age cohort, as well as the uncertain future of the priesthood, may also account for the drastic decline of seminarians. Unless this decline is reversed, in years to come the priesthood will be smaller and older in its average age. Both the many well-publicized resignations from the priesthood and the general decline in organizational practice resulting from the encyclical letter *Humanae Vitae* have also contributed to the decline in the number of young men who choose the priesthood as a profession.

The priesthood, then, is in better shape than its public image and stands solidly in support of change in the church and solidly against the old sexual morality. But like the rest of institutional Catholicism, it is caught in the present crisis of leadership.

The severity of this crisis can be judged by comparing bishops with diocesan priests, and the religious superiors with order priests.† Most religious superiors are elected by their priests and are of approximately the same age and attitude as their confreres. Bishops, however, are appointed directly by Rome and are very different in attitude, age, and values from their priests. Thus 70 percent of the diocesan priests wish to see bishops elected by priests, but only 24 percent of the bishops do so. Seventy percent want to see the bishops themselves participate in the election of the pope, but only 43 percent of the bishops want the same privilege of participation. Seventy-eight percent of the diocesan priests and 58 percent of the bishops would support some kind of due process appeals system in the church. Sixty-four percent of the priests believe in small informal parishes, as do only 32 percent of the bishops. Sixty-three percent of the diocesan priests are in favor of optional celibacy, but only 11 percent of their bishops agree with them. Eighty-three percent of the bishops and 30 percent of the clergy think that the faithful are bound to avoid all methods of artificial contraception. Sixty percent of the bishops and 26 percent of the diocesan priests believe that absolution should be denied to those who refuse to avoid the use of contraceptives. Seventy-two percent of the bishops thought in 1970 that the issuance of *Humanae Vitae* was a competent and appropriate

---

* Which led one NORC colleague to remark that the typical American Catholic knew lots of priests but not many college professors.

† Between 1965 and 1975 there was a decline of 25 percent, from 180,000 to 135,000, in the numbers of religious sisters; and a decline of 30 percent, from 12,000 to 8,600, in the number of religious brothers. No adequate national sample research has been done on either of these bodies—though one large but inept survey of the sisters was attempted in the middle 1960s.

use of papal teaching authority, but only 36 percent of their priests agreed. Fifteen percent of the bishops and 55 percent of the diocesan priests thought that the church could grant divorces even in sacramental marriages. Fifty-five percent of the bishops and 29 percent of the priests thought that in most cases deliberate masturbation is a mortal sin.

The striking differences in attitudes and values between bishops and their priests is *not* a function of their different ages; bishops are more conservative than priests of their own age groups. It is precisely the difference in attitudes which may be the crucial institutional flaw in American Catholicism. The priests, by and large, reflect the difficulties, aspirations, and problems of their people, but these insights are not reflected upward to the bishops, or at least not clearly or strongly enough for the bishops to reflect the attitudes and values of the clergy. In the absence of research (and the NORC project on the priesthood was virtually disowned as soon as its unpleasant findings on the encyclical *Humanae Vitae* were discovered), or shared values with their clergy and the laity, bishops will report to Rome not what is happening within the American church institution but what they think is happening, which is, as with most humans, not very different from what they think ought to happen or what they think the central authority wants to hear.

And given the way they are selected, there is no reason why one would expect bishops to reflect the feelings of their clergy or their people. Since the appointment of the first apostolic delegate in 1895, the recommendations of this temporary resident in Washington, D.C. have been of decisive importance. His recommendations were usually made after consultation with a group of ecclesiastical kingmakers—frequently themselves creatures of the delegate or his predecessor. Cautious, conservative, and "safe" men were usually appointed to the largest and most influential cities in the country. These safe men could be counted on to do nothing creative or imaginative, nothing to make any waves, nothing to change the status quo. Timid, careful, frequently not very intelligent, and almost never with any advance training besides some in canon law, these leaders were the kind of men who would prevent the resurgence of the "Americanism" of the late nineteenth century that disturbed Rome so much; they would also keep ecclesiastical revenues flowing from the United States to Rome (and also into the personal coffers of the apostolic delegate).

The process has not worked perfectly. At the Second Vatican Council, a number of distinguished American leaders emerged—notably Ritter of St. Louis, Meyer of Chicago, and Hallanan of Atlanta. Unfortunately, these brilliant and effective leaders were all dead shortly after the conclusion of the Council (in the case of Meyer, while the Council was still going on). They were replaced by incompetent nonentities. Indeed, in the years after the Council a systematic attempt was made by the Roman Curia to punish the American hierarchy for its support of ecclesiastical reform measures during the Council sessions. No one who had taken a vigorous public stand was promoted, and only those who could be counted on to resist change at all costs received appointments to major archdioceses. The result of this policy was that American Catholicism faced the crisis of the late 1960s and early 1970s with the least distinguished bunch of bishops of its whole history, as well as a collection of archbishops which, with one or two exceptions, ranged from the dull to the psychopathic. Such men could hardly be looked to for leadership out of the swamp—especially since most of them never knew they were mired in one.

In 1971 Archbishop Benelli, the assistant secretary of state and defacto prime minister of Paul VI, made a visit to the United States during which he was reportedly horrified at the collapse he observed and shocked by the ineptitude of Catholic leadership. A few years later a new apostolic delegate, a Belgian veteran of Third World assignments named Jean Jadot, was appointed. Jadot had not been trained to be a papal diplomat; he was, in fact, the first non-Italian to be sent to the apostolic delegation. Jadot has been responsible for a number of brilliant appointments to major dioceses and for restoring some measure of confidence among those clergy who pay any attention to what the apostolic delegate is doing. At the same time as Jadot's appointment, Joseph Bernardin, the flexible, pragmatic, and sophisticated archbishop of Cincinnati, was elected to the presidency of the American hierarchy. The influence of Jadot and Bernardin in Rome is uncertain; at the present time their enemies are campaigning vigorously against them. It is a sad commentary on the rapid deterioration of institutional Catholicism in the United States to say that only Jadot and Bernardin represent effective forces for reversing the deterioration. Whether they will succeed remains to be seen; the odds against them are high.

The bishops also differ in some important demographic ways from their clergy. In 1970, 68 percent of the bishops were over fifty-five, as compared with 23 percent of the priests. Forty-two percent of the bishops came from cities over one-half million in size, as did 34 percent of the priests. Half of the bishops are Irish, as are 34 percent of the priests and 17 percent of the Catholic population. However, there was no difference in the occupational prestige score of the fathers, and compared with priests of the same age level, bishops did not come from better-educated families. The demographic differences do not seem to be very important save for the overrepresentation of the Irish in the hierarchy (corresponding to American secular political life). The fundamental differences between bishops and priests seem to be those of perspective, which appear to be largely the result of the way bishops are selected.

The warning of John Carroll, the first bishop of the United States, that American Catholicism would be in serious trouble if its leadership was selected by any means other than election seems to have come true —even though the fulfillment of the prophecy took 200 years.

How do the Catholic laity respond to their clergy? One could say with some appropriateness that the clergy are like Willy Loman—liked but not well liked, admired but not altogether respected. However, the principal reason for decline in support for the priesthood as a vocation for one's son seems to be closely linked to the general deterioration of institutional Catholicism in the United States, which itself is closely linked to the birth control stand of the church and confidence in the papacy.

American Catholics give their clergy rather low marks on professional performance (Table 8.1). Fewer than half considered them to be very understanding in dealing with the personal problems of adults or young people. Only 20 percent consider their sermons to be "excellent." The first and third items involving professional competence were asked in both the 1952 and 1965 *Catholic Digest* studies of American religious attitudes.[4] This enables us to trace a steady deterioration in the Catholic evaluation of the professional performance of their clergy, a deterioration not matched by Protestant and Jewish evaluations. In 1952

TABLE 8.1
Attitudes of Catholics toward their Religious Leadership

A. PROFESSIONAL COMPETENCE (PERCENT AGREE)	PERCENT
Priests are "very understanding" of the practical problems of their parishioners	48
Priests are "very understanding" of the problems of their teen-aged parishioners	47
Priests' sermons are of "excellent" quality	23

B. ANTICLERICAL ATTITUDES (PERCENT AGREE STRONGLY OR SOMEWHAT)	
Priests expect laity to be followers	43
Priests are not as religious as they used to be	47
Priests are unconcerned about people, only themselves	17

C. POLITICAL ACTIVISM (PERCENT AGREE STRONGLY OR SOMEWHAT)	
Priests should not use pulpit for social issues	51
Priests may get involved in national and local politics	48

D. VOCATIONS	
Unhappy if daughter became a nun (percent disagree strongly)	50
Very pleased if son became a priest (percent agree strongly)	50

Catholics rated their clergy higher than did Protestants or Jews in their preaching performance and their ability to handle problems, while in 1965 they rated them lower than did Protestants and Jews. Unfortunately, we do not have any comparative data for Protestants and Jews in 1974. Whether there actually has been a decline in professional competence among the clergy or an increase in expectations among the Catholic laity is very difficult to judge. It should be noted, though, that the decline in positive evaluation is not limited to the well educated; it has occurred at all educational levels.

While only a minority of the Catholic population gives the clergy high professional ratings, a majority reject charges that the clergy are not as religious as they used to be and that they expect the laity to be followers. Only 17 percent can be persuaded that "priests have lost interest in the problems of the people and are concerned only about themselves."

On the subject of political involvement, the Catholic population divides evenly, with half saying the priests should not use the pulpit to discuss social issues and about half saying that it is all right for the priest to get involved in national and local politics. The picture that emerges is a vote against professional competence but no hint of anticlericalism, with a split decision concerning political involvement.

We observed earlier that only half of the American Catholics would be "very pleased" if their son decided to become a priest. The same proportion reject strongly the idea that they would be unhappy if a child of theirs should choose to be a nun (a decline of 10 percentage points since 1963). The priesthood and the religious life for women seems to have about the same level of approval among Catholics in 1974.

Finally, 29 percent are in favor of the ordination of women priests. Thirty-two percent express a great deal of sympathy with those who have left the priesthood (and 40 percent more express some sympathy). Eighty percent could accept a married clergy, and 62 percent are positively in favor of a married clergy.

In summary, Catholic support for a religious vocation for their children has declined in the last ten years; there is a fair amount of sympathy with those who have left the priesthood, and strong support for a married clergy. Priests are rated relatively low on professional ability, but the overwhelming majority do not think they are selfish or unconcerned about the people. The population divides evenly on the controversial issue of the social and political involvement of priests.

As I noted in Chapter 7, there has been a positive correlation between attendance at Catholic schools and support for a priestly vocation for one's son; in 1974 this relationship became negative for Irish Catholics. If the change in sexual ethics had an especially deleterious effect on those Irish who had Catholic education, one would expect that those who had a Catholic education would be significantly fewer in 1974 than the rest of the Irish if they were now in disagreement with the church's sexual ethic. Those Irish who had turned against the church's sexual morality from the Catholic education would have turned even more sharply against support for a priestly vocation than the typical Catholic respondent. This was the case. In 1963, 83 percent of the parochial school-educated Irish were ready to say that they would be very pleased if their son would choose the priesthood—17 percentage points above the national average. By 1974 this percentage had fallen to 45—5 percentage points below the national average. Almost all of the change was in that large group of Catholic-educated Irish who reject the church's sexual teaching. This shift of attitude seems to have had a particularly harmful effect on attitudes toward the clerical vocation among those who were its strongest supporters—the Catholic-educated Irish.

The Catholic clergy are in better personal condition than they are in public image. Although resignations from the priesthood continue (and probably will continue now that it is an option for those who are dissatisfied), the morale, satisfaction, and personal maturity of the clergy compare favorably with that of other men of the same age and with the same educational background. The clergy, unlike the hierarchy, are in contact with the problems of their laity. The laity, while they like their clergy and think they are working hard, are not impressed with their professional competence, and are substantially less likely to support a priestly vocation for their sons. However, the NORC analysis indicated that the reason for the decline of support for vocation was not the relatively low levels of respect for professional competence but the more generalized turning away from the institutional church caused by the changing sexual attitudes of Catholics and by the disaffection caused by the encyclical letter *Humanae Vitae*. Ironically, the clergy, who are probably suffering a loss of recruits because of the letter, are as much if not more opposed to it than the laity.*

---

* Recent research by Richard Schoenherr updates the 1970 NORC study and indicates that there has been a very slight decline in the annual resignation rate, and that in the last decade some 11 percent of the diocesan clergy in the country have resigned. Losses to the priesthood through death and resignation have barely been replaced by ordination. According to Schoenherr, the 9 percent retirement of older clergy represents the approximate net decline of diocesan clergy at work in American dioceses (as opposed to the foreign missions) since 1965. Schoenherr predicts that the combination of resignation (disproportionately likely among the more liberal clergy) and the decline in vocations will produce an aging and more conservative clergy in years to come.

REFERENCES

1. "American Priests," A Report of the National Opinion Research Center prepared for the United States Catholic Conference, mimeographed (Chicago: NORC, 1971).

2. Norman M. Bradburn, *The Structure of Psychological Well- Being* (Chicago: Aldine, 1969).

3. Everett L. Shostrom, *EITS Manual for the Personality Orientation Inventory (POI): An Inventory for the Measurement of Self-Actualization* (San Diego, Calif.: Educational and Industrial Testing Service, 1966).

4. The Ben Gaffin *Catholic Digest* studies are reported in John L. Thomas, S.J., *Religion and the American People* (Westminster, Md.: Newman Press, 1963).

CHAPTER

9

Catholics and
Their Schools

CATHOLIC SCHOOLS have always been offensive to many non-Catholics. They were prime targets for nativist bigotry; long ago, in some cities, schools were burned to the ground. Even after the passing of the most virulent forms of nativism, subtle and more sophisticated bigots still resented Catholic schools. They were "un-American," no one knew what went on in them, their teachers were not adequately prepared, the subject matter taught was not academically sound, the graduates of Catholic schools had an inferior education. Most serious of all, Catholics were separated from other Americans by the "divisive" parochial school system; hence, the schools both impeded the assimilation of immigrants into American society and set up barriers to understanding among the different groups within the society. Catholic schools have always been a matter for conflict, as Catholics have (unreasonably) tried to obtain state support for their separate school system. Catholic schools, it was suspected, were not the result of the free choice of Catholic parents but rather of the constraints of church discipline (as Mr. Justice Powell recently asserted). It was hoped that as Catholics became more Americanized, they would recognize the inferiority and divisiveness of their schools and turn away from them. One writer even suggested that as a price for ecumenical dialogue, Catholics should be ready to abandon not only their plea for state and federal aid for parochial schools, but also to abandon the schools themselves.

A number of points were overlooked in such characterizations of the parochial school system. The United States is virtually the only country in the North Atlantic world that does not accept Catholic schools as an integral part of its educational system. Catholic schools make a major contribution to education in northern urban areas, and indeed, the Catholic school system in Chicago is the fourth largest primary and secondary school system in the country (just behind the public school systems of New York, Los Angeles, and Chicago). More recently, Catholic schools have acted as an alternative educational opportunity for tens of thousands of inner-city blacks—a service which is almost completely ignored in discussion of the problems of equality of educational opportunity for blacks. Finally, as private schools have proliferated in the South and the North as a response to the inadequacy or racial integration of urban public schools, the argument that "everyone" sends his or her children to public schools has lost much of its effectiveness. "Everyone" doesn't send the children to public schools; southern racists, Harvard professors, and members of Congress and the government bureaucracy in Washington are very likely to choose private education for their children.

Colin Greer has contended that public schools have failed the blacks and the Latins, but that this failure to respond to the needs of immigrants is merely part of the history of public education, and that neither did public schools respond to the needs of earlier immigrants.[1] Greer considers the Catholic example of setting up separate schools to deal with the needs of immigrant children to be an example that blacks and other minorities in the big cities might well imitate.

Paradoxically, just at the time when scholars like Greer are engaged in a sympathetic reappraisal of Catholic schools, there has been a serious crisis in the schools. The shortage of nuns has led to a dramatic increase of lay teachers and a consonant raising of costs. Priests and sisters are no longer as confident as they were that the apostolate of Catholic education is a valid vocation. Attendance in Catholic schools has declined, in substantial part because Catholic school construction has almost ceased since the Vatican Council. There was controversy within the Roman church before 1963 about the existence of a separate Catholic school system, but in the past decade this controversy has risen to a crescendo. It is safe to say that only a handful of Catholic theoreticians are prepared to defend the continuation of Catholic schools.

Data published in the annual *Official Catholic Directory* is generally of a quality to give professional statisticians nightmares (it underestimates Catholics by between 2 and 3 percentage points of the American population, for example—which represents between four and five million people), but it does give some idea of the rise and fall of

Catholic schools since the end of World War II. In 1945 there were 10,912 Catholic schools with 2,590,660 students, according to the *Directory*. In 1965 the schools had increased by 31 percent (to 14,296) and the students by 35 percent (to 3,505,186). But in the most recent *Directory* (1975) the schools had decreased 24 percent and enrollment had fallen by 35 percent.[2]

Elementary school enrollment dropped by slightly under 120,000 between 1974 and 1975—a 3 percent decline for that year, a total 7 percent decline since 1973. On the other hand, enrollment in Catholic high schools in 1975 reached almost the one million mark it attained in 1965—an increase of 13,638 over the previous year. Enrollment in Catholic colleges also increased by more than 15,000. It is now 422,243 —almost 40,000 higher than it was ten years ago. In 1965, 14 percent of the grammar school children in the country were in Catholic schools; in 1975 this figure fell to 8 percent. The secondary school drop was from 9 percent to 7 percent.

If one speaks of a decline in Catholic school attendance, then, one must be careful to specify that most of this decline has taken place at the elementary school level. Enrollment did fall at the secondary school level, but it has begun to inch back up toward its 1965 high. College enrollment is actually higher now than it was ten years ago, and seems to be continuing to climb despite all the problems that private higher education is experiencing.*

When the decline in Catholic school enrollment became public knowledge, articles began to appear in the national media in which certain Catholic "experts" (such as John Deedy's periodic writings in the *New York Times*) endeavored to explain the phenomenon. Not surprisingly, the experts offered the kinds of explanations the readers of such journals wanted to hear: the costs of Catholic schools were rising rapidly; Catholics, particularly in the suburban areas, were coming to understand that public education was better academically; Catholics were now sophisticated enough to realize that religious values were much better inculcated in the homes than in the schools; and the diminishing fervor of Catholics in the wake of the Second Vatican Council led them to be suspicious of the need for any religious education. The long-awaited demise of Catholic schools, it seems, was finally occurring.

The National Opinion Research Center (NORC) has conducted two

* All of the statistics here should be viewed with caution. The U.S. government (see *Current Population Reports*, February 1975) reports on private school enrollment. Comparing the Catholic school students (as cited in the *Catholic Directory*) with the total private school enrollment in 1965 reported by the U.S., one finds that Catholics were 98 percent of the private school elementary students. In 1975 they were 82 percent. I doubt that there has been that much of an increase in private non-Catholic school enrollment. At the secondary level, the proportion at both time points was 78 percent. I suspect that at one time or other the Catholic data were in error.

national sample surveys of American Catholics, one in 1963, the other in 1974.[3] The purpose of the first study was to measure the effect on adult attitudes and behaviors of having attended Catholic schools when one was of school age. We wanted to learn whether the schools were socially divisive or economically disadvantageous, and to determine what attitudes were held toward parochial education. The second study was conducted to determine what the changes in the Catholic church since 1963 had done both to the results of the Catholic school attendance and attitudes toward the schools.

There was not the slightest evidence to justify the contention that parochial schools were being used to escape racial integration—especially since three-fifths of the parents with children in Catholic school reported that the schools were integrated. The principal advantage of Catholic schools in both 1963 and 1974 (mentioned by about half the respondents) was religious instruction. One-third of the respondents listed better discipline and one-quarter mentioned better education as being among the principal advantages of Catholic schools. In both 1963 and 1974, the main reasons (listed by half the respondents) for not sending children to Catholic schools was the simple fact that schools were not available.

In 1963, 44 percent of the Catholic children of primary and secondary age were in parochial schools; by 1974 the proportion had dropped to 29 percent—a decline of one-third. However, if one decomposes the 15 percentage point decline, one discovers that 11 percentage points of those 15 are concentrated in suburban areas. Nine points of the percentage difference can be accounted for by the increase of the "unavailability" of schools in the suburbs (Table 9.1). When one adds an additional percentage point accounted for by an increase in unavailability of Catholic schools in rural areas, we can see that two-thirds of the decline of Catholic school attendance can be accounted for by an increase of their unavailability. The rest of the change is the result of the increase in the cost of Catholic schools. In fact, the proportion arguing that there is better education in public schools (as a reason for nonattendance) has decreased in the last decade. Hence, there are only two reasons for the decline in Catholic school attendance: unavailability of new schools in the suburban areas to which the Catholic population has moved in the last decade, and increased cost of attendance at schools. The other reasons cited by the Catholic "experts" simply are not true, and indeed the allegedly superior education of the public schools is less important in explaining nonattendance now than it was ten years ago.

Nor is there any evidence of a decline in support for the principal of Catholic education among American Catholics. In a number of different local, regional, and national surveys taken in the last twenty

TABLE 9.1

Reasons for Children Not Going to Catholic Schools, 1963 and 1974*

REASONS	CITY		SUBURB		RURAL		TOTAL		DIFFERENCES BETWEEN 1963 AND 1974†			
	1963	1974	1963	1974	1963	1974	1963	1974	CITY	SUBURB	RURAL	TOTAL
Public schools better	03	02	04	03	03	02	10	07	−01	−01	−01	−03
Costs	04	07	04	07	02	04	10	18	03	03	02	08
Unavailability of school	03	03	11	20	08	09	22	32	00	09	01	10
Other	02	02	10	10	02	02	14	14	00	00	00	00
Total not attending	12	14	29	40	15	17	56	71	02	11	02	15

SOURCE: Greeley, McCready, and McCourt, *Catholic Schools in a Declining Church*, p. 236.

*1963 percentage not attending = 56; 1974 percentage not attending = 71

† Total differences = 15

years, and in response to a number of different question-wordings, at least three-quarters of the Catholics in the United States have endorsed the principal of Catholic schools, and there is simply no evidence that the level of endorsement is falling. In the 1974 NORC survey respondents were presented with an item which said, "The Catholic school system has outlived its usefulness and is no longer needed in modern-day life." Only 10 percent of the respondents endorsed the item; 65 percent rejected it strongly, and another 24 percent rejected it "somewhat." Furthermore, there was no age correlation with responses to this item. Catholics in their twenties and thirties were as likely to agree with it as those in their forties, fifties, and sixties. Sixty-six percent thought that lay teachers could do as good a job as nuns did in the Catholic schools, 75 percent rejected the idea that parents who send their children to Catholic schools would settle for lower academic standards, 76 percent endorsed federal aid to Catholic schools, 62 percent endorsed a tax rebate for parents of parochial school children, and 40 percent thought that there would be federal aid for Catholic schools were it not for anti-Catholic feelings in the government.

Catholics indicated they were willing to use their money in support for Catholic schools. When asked whether they would increase their contribution to the Sunday collections if it was necessary to keep the local parochial school going, 80 percent of the Catholic respondents replied that they would. Fifty percent said they would give more than $50 a year in additional contribution, and 25 percent said they would give more than $100 a year in additional contribution. Again there was no correlation with age; young Catholics were as likely as older Catholics to say they would contribute additional money to keep the local school open. William McCready has estimated that there is approximately $1.8 billion of potential support available for Catholic schools that has not been tapped by Catholic leadership.[4] (This is exactly the same amount of money that represents the decline in Catholic church contributions in the last decade, as reported in Chapter 7.)

Despite the lower levels of attendance in Catholic schools, there is still vigorous support for them among the American Catholic population, as well as a willingness to back up this support with a higher level of financial contributions. Why, then, are some schools closing, and why are schools not being built in the suburbs where Catholics now live?

One must understand the peculiar decision-making process in the American Catholic church. Bishops rarely consult with priests or laymen in making major decisions in their dioceses, and parish priests have begun only recently to consult with their parishioners. Even when consultation does take place, both bishops and parish priests feel free to make whatever decision they think best, regardless of the opinions of their subordinates. The leadership of the American church has only two important reference groups in its decision making: its fellow leaders and ecclesiastical authority in Rome. American bishops, with some exceptions, listen to their fellow bishops when it comes to making decisions, and not to their priests or laity. As we noted in Chapter 8, most of the bishops appointed since the Second Vatican Council have been incompetent, inept, and mediocre. Such men are likely to be obsessed by the details of organizational administration, and particularly by the complexities of ecclesiastical finance—an area about which they know little but from which they fear much. Noting the skyrocketing costs of Catholic schools and reading in the newspapers and magazines that Catholics were no longer interested in continuing Catholic education, many bishops became frightened at the possibility of financial disaster if they continued to divert resources into Catholic schools. Apparently they were unaware of the tremendous increase in real income that has occurred among American Catholics in the last twenty years, and their willingness to continue to make notable financial sacrifices for the parochial schools. In the closed communication network of the American hierarchy, a decision emerged, more or less implicitly, to deemphasize Catholic school construction while continuing to profess strong loyalty and support for Catholic schools. Attempts were made—usually not very skillful—to get federal or state support, but these attempts ran afoul of the Supreme Court's determination to prevent aid to Catholic schools. The bishops thus had a villain to blame—the government and the courts—and could continue to close some schools and neglect the building of new ones while endorsing them in principal. Failure to build new schools despite demand for them and willingness to pay the costs involved results from a combination of incompetence, fear, and pluralistic ignorance. The laity are used to having little impact on the ecclesiastical decision-making process, and they have no institutions available through which they can make their collective wishes known to ecclesiastical leaders.

Those unfamiliar with the internal decision-making processes of the church and unaware of the monumental incompetence of the present generation of bishops (again, with some exceptions) will be astonished that Catholic schools can be phased out despite the wishes of the Catholic population. (In the Archdiocese of Chicago, for example, no new Catholic schools have been built in the last six years, even though the diocese is reputed to be the world's richest. Many millions of dollars have been put into a gimmicky closed-circuit television network which no one watches. Both the decisions to stop school construction and to pour money into the television network were made solely by the archbishop.) Surely, the reader may ask, the bishops and the school administrators were aware of the research evidence that showed the deep support for Catholic schools in the Catholic population? The reader can begin to understand the institutional problems of American Catholicism when he comprehends the fact that bishops and school administrators couldn't care less about research evidence when it comes to decision making. They do not believe research findings. In fact, they do not believe in research; they are much more worried about what their fellow bishops would say or what Rome would think than they are about the wishes and expectations of the rank and file laity.

But is the continued enthusiasm of Catholics for their schools justified? Do Catholic schools produce an outcome that is commensurate with the time, resources, and energy expended on them? It is difficult to answer these questions because there is no clear evidence of what the outcome would be of other efforts. However, two things can be said: (1) The various forms of religious instruction outside the Catholic school context have almost no impact on adult religious attitudes and behaviors; and (2) attendance at Catholic schools does produce a measurable impact on adult religious attitudes and behaviors over and above the influence of the family.

To summarize the findings of NORC's two research reports, a "Catholicity" factor was developed. It is a summary measure of church attendance, personal devotion, organizational involvement, financial contribution, and doctrinal and ethical orthodoxy. This factor was then analyzed in a mathematical model which produced "betas" or "standardized coefficients" for parental religiousness, age, sex, educational achievement, number of years of Catholic education, and church attendance for one's spouse. The standardized coefficients are "net relationships"; thus the coefficients between age and Catholicity (.05 in 1963 and .20 in 1974) represent the "pure" influence of age, with parental religiousness, sex, educational achievement, years of Catholic education, and spouse's church attendance held constant. The effects of all other variables are filtered out in order to get the standardized coefficient. Thus the relationship between years of Catholic education (.15 in 1963 and .23 in 1974) is the pure effect of Catholic schools, filtering out the impact of all the other variables. Parental religiousness, in particular, is *not* mixed in with the standardized coefficient for years of Catholic education.

As we can see in Tables 9.2A and 9.2B, the model explains 9 percent more of the variance in 1974 than it did in 1963. The predictive power of age went up from .05 to .20, and the predictive power of sex went down from .28 to .19. Education, once a moderately strong predictor of Catholicity at .15, now has a negative relationship with that variable (— .07). Parental religiousness has declined, while spouse's church attendance has moved from .42 to .50. Catholic education was .15 and grew to .23; it became a stronger net predictor of one's position on this scale than age, sex, educational attainment, and parental

TABLE 9.2A

Average Standardized Coefficients with "Catholicity" Factor in Religious Behavior Model, 1963 and 1974

	1963	1974
Parental religiousness	.28	.22
Age	.05	.20
Sex	.28	.19
Educational achievement	.15	−.07
Years of Catholic education	.15	.23
Spouse's church attendance	.42	.50
Total variance explained	.38	.47

SOURCE, tables 9.2A and B: Greeley, McCready, and McCourt, *Catholic Schools in a Declining Church*, p. 188.

TABLE 9.2B

Standardized Coefficients (Betas) Between Years of Catholic Education and "Catholicity" Factor for Certain Subgroups

	1963	1974
Male	.18	.23
Female	.17	.15
Religious parents	.19	.23
Less religious parents	.03	.03
Under 30	.08	.21
Over 30	.16	.12
Irish	.06	.05
German	.12	.10
Polish	.08	.27
Italian	.13	.25
Spanish-speaking	.07	.24

religiousness. Dramatic change in the church, in other words, has affected the importance of other variables in determining adult religiousness, and has increased the impact of Catholic education.

The number of years one has attended Catholic schools has declined somewhat as a predictor of Catholicity for women, but the relationship between Catholic education and Catholicity has risen for men and gone up for all respondents from very religious families. The result is that the differences between men and women and between religious and less religious families increased over the decade covered by the studies. It rose to .08 between men and women, as opposed to .01 at the beginning of the decade, and to .20 between those from less religious and those from more religious families, as opposed to .16.

The sharpest differences, however, have occurred in the age groups. There was only a .08 beta between years of Catholic education and Catholicity for those under thirty in 1963 (the Cold War cohort); for their successors (the Vietnam generation) the relationship has jumped to .21. For those over thirty, the beta has declined from .16 to .12. There has been a change of .17 in the differences of impact of Catholic education on adult religiousness in the two age groups. There has also been a decline in the relationship for the Germans and Irish. Large increases occurred for the Italians, Poles, and Spanish-speaking.

So in the ten years covered by our surveys Catholic religiousness has gone down, but the importance of Catholic education has gone up because of its heightened impact on certain ethnic groups, young people, and men—the last two groups being those which show the sharpest decline in religiousness. In other words, Catholic education has slowed the decline more effectively for those groups among which the decline has been the most serious. Catholic education seems much less effective for those among whom the decline has been less serious. Furthermore, its relative impact is even greater now on those who will build the future —those from more religious families, young people, and men (who are the decisive religious socializers in families).[5] Thus, in terms of both cutting losses and restructuring for the future, Catholic schools seem substantially more important today than they were a decade ago.

The standardized coefficient between number of years in Catholic school and adult religiousness is not large (.23). How effective can Catholic schools be considered, with such a standardized coefficient? To a considerable extent the answer depends upon how much one believes adult behavior to be shaped by background influences, and how possible it is to find pertinent background variables. One scarcely can expect any background influence to produce an overwhelming correlation, for such an influence would take randomness, variety, and complexity out of human action and personality. All we can say about our religious models presented here is that they are generally about as successful in predicting adult religiousness as are the traditional social stratification models of other social scientists in predicting income levels.

Catholic education in 1974 is as strong a predictor, on the average, of adult religious behavior as the religiousness of parents. It is stronger than one's level of educational attainment and sex. Furthermore, the average influence of Catholic schools remained constant between 1963 and 1974, while parents' religiousness, educational attainment, and sex declined in predictive effectiveness. (Age and spouse's religiousness increased, however.)

At the time the first NORC study was reported,[6] there was little material available from other educational impact research by which comparisons could be made. Since then, James Coleman and his colleagues studied the equality of educational opportunity and focused on the impact of education on blacks and whites.[7] Tables 9.3A and 9.3B originated in an analysis of the Coleman study that was prepared by Marshal S. Smith.[8] The standardized coefficients between per pupil expenditure in a school system and verbal achievement scores are virtually nonexistent for both blacks and whites, and the standardized coefficients between proportion white in a school and the academic achievement of blacks is notable (approximately .2) for only ninth-grade northern blacks. Generally speaking, the only substantial beta for family background variables with individual verbal achievement is parental education. Comparisons between the standardized coefficients of this study and those of the NORC study are only illustrative. The dependent variable of the Equality of Educational Opportunities Study (EEOS) is present performance in school, while the dependent variable in our

TABLE 9.3A

*Raw and Standardized Correlations Between per Pupil Expenditure,
Proportion School White, and Verbal Achievement Scores*

	PER PUPIL EXPENDITURE			PROPORTION WHITE		
	RAW	BETA 1*	BETA 2†	RAW	BETA 1*	BETA 2†
Sixth grade						
Northern blacks	.07	.01	.02	.15	−.01	−.03
Northern whites	.04	.01	.02	.16	.06	.04
Ninth grade						
Northern blacks	.00	.01	.04	.14	.22	.20
Northern whites	.07	.06	.08	.12	.06	.05
Twelfth grade						
Northern blacks	.00	.00	.01	.15	.09	.06
Northern whites	.08	.03	.04	.06	.06	.03

SOURCE: Adapted from Marshal S. Smith, "The Basic Findings Reconsidered," in
On Equality of Educational Opportunity, ed. Frederick Mosteller and Daniel P.
Moynihan, pp. 331-334. (This and following table are reprinted by permission of
the publisher, Random House, Inc. — © 1972.)

*Net of family background factor, facilities and curriculum factor, student body
factor.

†Net of all but family background factor.

research is adult performance after one has left school. But a comparison
does suggest that in social research, one must be content with relatively
modest results.

Whether the massive modern resources poured into Catholic
schools are worth the effort is a value judgment beyond the scope of
our discussion. Schools apparently do have as much impact on adult

TABLE 9.3B

Regression Coefficients for School Achievement

	NORTHERN BLACKS			NORTHERN WHITES		
	6TH GRADE	9TH GRADE	12TH GRADE	6TH GRADE	9TH GRADE	12TH GRADE
Reading material in home	.11*	.08*	.04	.12*	.11*	.17*
Items in home	.13*	.05	.02	.08*	.04	−.00
Siblings	.08*	.13*	.09*	.10*	.14*	.09*
Structural integrity of family	.06	.07*	.04	.06*	.08*	.03
Parents' education	.10*	.09*	.14*	.19*	.23*	.19*
Urbanism of background	.01	.04	.05	−.09*	.07	.01

SOURCE: Smith, "The Basic Findings Reconsidered," p. 266.

NOTE: Standardized regression coefficients for six background variables at grades 6, 9,
and 12 for northern whites and blacks. Each coefficient is estimated in an equation containing
all other background variables, *proportion whites in the school*, and three sets of schoolwide
variables measuring curriculum and facilities, teacher characteristics, and student body characteristics. Individual verbal achievement is the dependent variable.

*An asterisk next to a value signifies that this coefficient is significant at the .05 level
($t > 1.96$, $N = 1,000$).

religiousness as the religiousness of one's parents, a fact which might be considered not unimportant. Administrators of the Catholic schools seem to have made the decision to jettison the parochial school system; they are closing down many of the existing schools and not building new ones. As we have seen in this chapter, however, the clientele of these schools appear to have made a different judgment. But no one has bothered to consult them.

One of the more interesting aspects of the impact of Catholic education on adult attitudes and behavior is that it seems to have little effect at the present time on sexual orthodoxy, church attendance, or sacramental reception, but it correlates strongly (in standardized co-efficients) with such things as Catholic organizational activism, church contribution, support for the Second Vatican Council, reading Catholic magazines, newspapers, and books, and support for priestly and religious vocations in one's family, as well as positive attitudes toward the clergy. In addition, for those from highly religious families, for the young, and for men, Catholic education correlates strongly with private prayer. In the present religious crisis in Catholicism, in other words, Catholic schools not only have their strongest impact on the ones most likely to keep the church going—young people, those from devout families, and men*—it also seems to affect those kinds of behavior which will be especially important if the church is to survive the present crisis—organizational activity, contribution, support for religious vocations, personal devotion, and positive attitudes toward the clergy.

Despite the decline in attendance at Catholic schools and the general decline in religiousness among American Catholics, then, Catholic schools seem more important for the church's continuation in the United States today than they did ten years ago.

But what of the charge that Catholic schools are "divisive" and "educationally inferior"? Do they impede social and occupational success for those who attend them? Do they stand in the way of the acculturation of the ethnic groups into American society? The materials presented in Chapter 6 on the racial and religious attitudes of Catholics who went to Catholic schools cast grave doubt on the contention that Catholic schools foster negative attitudes toward other groups of the society. Furthermore, regardless of whether or not they attended Catholic schools, three-quarters of American Catholics, in both 1963 and 1974, reported that their three best friends in adult life were Catholic. American Protestants from comparable areas report their closest friends to be co-religionists, too. So selecting friends from within one's own religious

* McCready's research on religious socialization, "Faith of Our Fathers: A Study of the Process of Religious Socialization" (Ph.D. diss., University of Illinois, Chicago Circle Campus, 1972), shows that religious behavior is transmitted across generational lines primarily and almost exclusively through the father.

group is an American phenomenon to which Catholic schools make no extra contribution.

Students in Catholic schools generally do well on standardized achievement tests. Indeed, in the inner cities of large metropolitan areas, they do much better than public school students. However, it is difficult to make precise comparisons because we cannot hold constant either the race or the socioeconomic background of parents. It is certainly true that the inner-city Catholic schools serve a clientele which is almost entirely black and mostly non-Catholic, and produce much higher test scores than public schools of similar areas. But the fact that the family chooses to pay the tuition (anywhere from $300 to $700 per year) for attendance at Catholic school is an indication of a higher level of parental concern about educational outcomes than one would find in the rest of the population.

Hence if one wishes to find out whether Catholic schools have impeded the adult performance of those who attended them, one must look at measures which only indirectly reflect the academic activities which went on in the school—educational attainment, occupational prestige, and income.

For those who thought that Catholicism would inhibit economic achievement, it was logical to assume that Catholic schools, which represent the quintessence of the "Catholic ethic," would inhibit economic achievement even more. The NORC study of 1963 showed, however, that parochial school attendance correlated positively with economic achievement. There was profit to be made by going to Catholic schools.

In our reanalysis of the 1963 data (Tables 9.4A and 9.4B) we found that those who spent at least ten years in Catholic schools had 13.6 years of education in 1963 and 14 years in 1974; this was more than a one-year advantage over those who did not go to Catholic schools at all. The finding holds even when we apply the standardization technique of taking into account father's education. In other words, those who attended Catholic schools still hold on to their educational advantage, an advantage that is not purely a function of the superior education of their fathers.

Those who had more than ten years in parochial schools in both 1963 and 1974 also have a lead of more than ten units in occupational prestige over those who had no Catholic education. They are more than approximately eight units over those who had one to ten years of Catholic schooling. In 1963 the income advantage over the former was more than $1,300; over the latter it was almost $900. In 1974 this advantage increased to $1,700 over those who had no Catholic education and almost $2,000 over those who had one to ten years. What seems to provide the economic payoff in 1974 is a lot of Catholic education. However, the difference in occupational prestige diminish consid-

TABLE 9.4A

Achievement of Catholics by Parochial School Attendance, 1963 and 1974
(For only those respondents with more than ten years education)

1963	FATHER'S EDUCATION (IN YEARS)	OWN EDUCATION (IN YEARS)	OCCUPATIONAL PRESTIGE SCORE	INCOME (IN DOLLARS)
0 years in Catholic school(365)	7.6	12.8	38.4	7,977
1-10 years in Catholic school(491)	8.6	13.0	42.9	8,526
More than 10 years in Catholic school(211)	9.3	13.6	51.0	9,300
1974				
0 years in Catholic school(201)	8.7	12.8	40.1	14,865
1-10 years in Catholic school(241)	9.7	12.8	40.2	14,701
More than 10 years in Catholic school(132)	9.9	14.0	47.9	16,628

SOURCE, tables 9.4A and B: Greeley, McCready, and McCourt, *Catholic Schools in a Declining Church*, pp. 196, 197.

TABLE 9.4B

New Educational, Occupational, and Income Difference From Those
With More Than 10 Years Catholic Education, 1963 and 1974
(For only those with more than ten years education)

1963	OWN EDUCATION		OCCUPATIONAL PRESTIGE		INCOME (IN DOLLARS)	
	GROSS	NET*	GROSS	NET[†]	GROSS	NET[‡]
0 years in Catholic schools	−0.8	−0.7	−12.6	−2.2	−1,323	− 476
1-10 years in Catholic schools	−0.6	−0.6	− 8.1	0.0	− 874	− 355
1974						
0 years in Catholic schools	−1.2	−1.0	−11.5	−4.3	−1,763	− 537
1-10 years in Catholic schools	−1.2	−1.0	− 7.9	0.0	−1,927	−1,091

*With father's education taken into account.

[†]With father's education and own education taken into account.

[‡]With father's education, own education, and occupation taken into account.

erably when we remove the effects of both the respondent's and respondent's father's educational level through standardization. At both points in time those who went to Catholic schools maintain their advantage over those who did not (though it becomes rather small). But there is no difference between the two Catholic school groups. However, the income differences remain at both points in time, and there is still more than $1,000 advantage in having more than ten years of education in Catholic schools.

There are a number of possible explanations for this economic advantage of Catholic education—an advantage which ran against the 1963 expectations:

1. Catholic schools are academically better than public schools, and provide their students with better intellectual skills, which in turn enable them to be more successful in the occupational world.
2. Those who attend Catholic schools learn habits of self-restraint and diligence (which is to say that they acquire the Protestant ethic?) that enable them to do better in the world of economic achievement.
3. Even if one holds constant the educational achievement of parents, it may still be the case that Catholic school attenders come from families where there is a higher motivation for economic achievement. The financial sacrifice made to provide Catholic education, for example, may be one evidence of such familial motivation.
4. Writers from the National Bureau of Economic Research (NBER) have recently raised the possibility of family influence within religious groups as an explanation of economic achievement:

Economists and other social scientists have recently begun to pay close attention to the possible role of preschool investments in children by parents, as it affects subsequent educational attainment. . . . Parental influences of this sort may also have effects on market productivity and earnings over and above any impact on school performance, and if so, returns on education can be affected.

To show the potential importance of these kinds of factors [possible influence on earnings of different amounts of parental time spent with preschool or school-age children], it is worth pointing out that cultural background as reflected by religious preference has a very powerful influence on observed earnings in both the Taubman-Wales and the Hause chapters. In the data sets used for both chapters, respondents were asked to report their religious preference as among Protestant, Catholic, Jewish, and other (including none). Taking account of family background factors like father's and mother's education and occupation, variables for both Jewish and Catholic religious preference have a significant (positive) impact on observed earnings relative to respondents reporting a Protestant preference. The Jewish religious preference variable also shows a significant impact on earnings in the Rogers sample.

Although the precise factors reflected in these religious preference variables are unknowns, plausible hypotheses are that they reflect differences in the cultural background to which respondents were exposed during their formative years, or differences in the quantity or quality of parental time inputs, rather than differences in specific religious values or practices. The appropriate research stance seems clear. The existence of strong statistical differences in behavior patterns associated with religious preference variables— or, as in other studies, with variables reflecting race or sex—points toward the existence of forces whose influence needs to be better understood and more fully interpreted, rather than toward an inference of causal relationships from observed statistical associations.[9]

If such family background factors do correlate with religion, then it would not be surprising to find such background dynamics especially vigorous in those families who were sufficiently Catholic to pay for ten or more years of Catholic education for their children.

Another possibility is that those who have considerable amounts of parochial schooling are in some fashion better integrated into the Catholic community, and for psychological or financial reasons this integration has a certain profitability. There are moderate correlations among income, education, and occupation and growing up in a Catholic neighborhood, living in a Catholic neighborhood now, and having a high proportion of your three closest friends Catholic (in excess of .1). However, there are almost no differences among the three educational groups (Table 9.5A) in their mean scores on these measures, so we cannot use the standardization model of holding those differences constant. There remains the possibility that even though the groups have on the average, for example, the same proportion of their three closest friends Catholic, they make different economic use of these friendships. They may be able to convert them differentially into income.

It must be remembered that the "conversion rate" in a regression equation (the "b") represents the change in the metric of the dependent variable which is accounted for by a change in the metric of the independent variable. Thus a "b" between income and proportion of friends Catholic would represent how much income one adds for each new

TABLE 9.5A

Deviations from Mean for Catholic Educational Groups in Catholic Community Measures, 1974

	GROWING UP IN CATHOLIC NEIGHBORHOOD	LIVING IN CATHOLIC NEIGHBORHOOD NOW	THREE BEST FRIENDS CATHOLIC
0 years in Catholic schools	.03	−.02	−.05
1-10 years in Catholic schools	−.01	.10	−.10
More than 10 years in Catholic schools	−.02	−.05	.27
Mean	(3.26)	(3.00)	(2.02)

SOURCE, tables 9.5A-C: Greeley, McCready, and McCourt, *Catholic Schools in a Declining Church*, pp. 200-202.

TABLE 9.5B

Conversion Rates ("B") for Living in Catholic Communities
by Catholic Educational Groups, 1974
(for those with ten or more years education)

	GROWING UP IN CATHOLIC NEIGHBORHOOD	LIVING IN CATHOLIC NEIGHBORHOOD NOW	THREE BEST FRIENDS CATHOLIC
0 years in Catholic schools	$ 467	$ 731	$ 859
1-10 years in Catholic schools	219	−868	310
More than 10 years in Catholic schools	660	1727	3681

unit of Catholic friend, holding constant all the other variables in the equation. One can see in Table 9.5B that there is a surprising difference in the ability of the three Catholic educational groups to convert "friendship" into income. Each extra Catholic friend is worth $3,681 for those with ten years of Catholic education, and $59 for those with no Catholic education. There is a net loss of $310 of income for each Catholic friend among those who had between one and ten years of Catholic education. There is a similar though smaller phenomenon at work for living in a Catholic neighborhood now and growing up in a Catholic neighborhood. The impact of Catholic friendship on income is illustrated with another statistic in Table 9.5C. The beta for those with all Catholic education is .41, for those with no Catholic education it is .07, and for those who were in between it declines to − .05. For those living in a Catholic neighborhood the figures are .26, .13, and − .15, respectively.

There are clearly a number of baffling mysteries at work. Why should there be any relationship between where you live and the religiousness of your friends, on the one hand, and economic success, on the other? And why should this relationship be positive for those who had a lot of Catholic education? Morris Rosenberg suggested some time ago[10] that growing up in a neighborhood where one was a minority had some effect on self-esteem, which, in its turn, weakened to some extent one's economic achievement. But our data indicate that the neighborhood in which you grew up is less important than the one you live in now and your present friendship patterns. Could it be that those who are in Catholic relational situations have better morale than Catholics who are not, and that therefore going to Catholic schools produces "better-adjusted" adults?

On Bradburn's psychological well-being scale ("How happy are

TABLE 9.5C

*Simple and Standardized Correlations with Income for Living in
Catholic Community Measures by Educational Groups, 1974
(for only those with more than ten years education)*

	GROWING UP IN CATHOLIC NEIGHBORHOOD		LIVING IN CATHOLIC NEIGHBORHOOD NOW		THREE BEST FRIENDS CATHOLIC	
	r	BETA	r	BETA	r	BETA
0 years in Catholic schools	.10	.06	.10	.13	.00	.17
1-10 years in Catholic schools	.01	−.03	.12	−.15	.02	−.05
More than 10 years in Catholic schools	.09	.07	.08	.26	.13	.41

you now? Very happy, pretty happy, not too happy"), we could find
no difference among the three Catholic educational groups in psychological well-being (Table 9.6).

But there were different relationships between proportion of three
closest friends Catholic and psychological well-being for the Catholic
educational groups (Table 9.6, part i), with those who had some Catholic

TABLE 9.6

Psychological Well-Being, Catholic Education, Catholic Friends, and Income, 1974

	0 YEARS IN CATHOLIC SCHOOLS	0-10 YEARS IN CATHOLIC SCHOOLS	MORE THAN 10 YEARS IN CATHOLIC SCHOOLS
i. Correlations between proportion of Catholic friends and psychological well-being for three Catholic educational groups	.10	−.18	.11
ii. Net correlation between psychological well-being and income with father's education, own education, occupation, proportion of Catholic friends held constant—for three Catholic educational groups (beta)	.28	−.08	.08
iii. Income conversion rates for proportion of three best friends Catholic with psychological well-being taken into account and not taken into account			
Well-being not taken into account	$ 859	$ −310	$3,681
Well-being taken into account	509	220	1,481

SOURCE: Greeley, McCready, and McCourt, *Catholic Schools in a Declining Church*, p. 203.

education displaying a negative relationship between Catholic friends and psychological well-being (— .18), and the other two groups showing a positive relationship. For those who had some Catholic education, a Catholic friendship network seemed to produce lower happiness for those who had either no Catholic education or a lot of it. Furthermore, with all the other variables in the model held constant (including number of close friends Catholic), there were positive relationships between income and psychological well-being for Catholics with no Catholic school education and those with a lot, and a negative relationship between psychological well-being and income for those with some Catholic education (Table 9.6, part ii). In this group, Catholic friends produce less happiness, and less happiness produces more income.

Finally, if one compares the conversion rates for Catholic friendship with psychological well-being in two ways, with psychological well-being included in the equation and with psychological well-being not included in the equation, one can see (Table 9.6, part iii) that while the differences are not eliminated, they diminish, and the negative relationship for those who had some Catholic education becomes positive. Thus, psychological well-being is involved in the differences among the three Catholic educational groups in their ability to convert friendship into income. Being integrated into the Catholic community does produce a higher level of psychological well-being, which in turn explains in part the higher economic convertibility of friendship. But why this should be true only for those who have more than ten years of Catholic education, and not for those who have less, remains a mystery. Furthermore, why it should also be true for those who have no Catholic education at all (though less dramatically than for those with more than ten years of Catholic schooling) is equally mysterious. Some of the differential convertibility of friendship seems to be related to social psychological factors, but much of it does not. It could certainly be the case that those who have had many years of Catholic education are tied into a Catholic business and commercial network which gives them an economic advantage; but then why would those who had no Catholic education also be able to profit from such a network? And why would the network be counterproductive for those who have had some Catholic education? There could easily be ethnic group differences at work, but our sample is not large enough to explore that possibility.

In summary, the advantage of parochial school attendance in educational attainment is not simply a function of the superior education of the respondents' fathers. The advantage in occupational prestige over those who had some Catholic school education vanishes between 1963 and 1974, but it continues over those who had no Catholic education. So, too, does the financial advantage, although in part that is the result of some aspects of Catholic community life we do not fully understand. The passage of ten years does not seem to have diminished these advantages of Catholic education in the slightest.

However one sorts out the intricate network of complex relationships through which we have just tried to thread our way, it certainly seems to be the case that neighborhood, friendship, psychological well-being, and Catholic education all interact in an intricate fashion to make Catholic school attendance a decided asset in the quest for educational, occupational, and economic success. As we shall see in subsequent chapters, Catholics tend to be more "communal" than their Protestant counterparts. ("Communal" is not used in the sense of "civic," as it is used in discussions of political style.) The intricate and dense network of human relationships in which one lives and works seems to have more effect on Catholic behavior than it does on Protestant behavior; Catholic friendship networks seem to be more tightly knit than Protestant friendship networks. In the absence of much more elaborate and detailed research than has ever been done on the subject—or is ever likely to be done—I will end with the observation that, for Catholics at least, communal reinforcement seems to strengthen occupational achievement and performance. It may well be that these communal strengths played a major role in the astonishing educational and economic achievement of the ethnic groups in the years since 1920. Juster and his colleagues rightly mention the importance of family in developing and supporting achievement orientations in the school and economic worlds,[11] but for Catholics it would appear that local community also plays a major role in strengthening and reinforcing upward mobility. Catholic schools apparently make a contribution to this mobility which is linked to communitarian ties but which has an independent impact that is specifically attributable to Catholic education. In any case, it is surely true that Catholic schools do not impede achievement during adult life. One cannot prove the inferiority of Catholic schools by appealing to the failure of the graduates of such schools in later years, for the graduates of Catholic schools are anything but educational, occupational, or economic failures.

Catholic schools also make a contribution to the adult social consciousness of those who attended them. Shirley Saldanha[12] has demonstrated that Catholic education in 1964 did not make a statistically significant contribution to the "social consciousness of adult Catholics" (Table 9.7) as measured on a scale combining their attitudes on race, anti-Semitism, social welfare, and charity. But a decade later the contribution of Catholic education had become statistically significant and not inconsiderable. (The beta in the table represents the "net" impact of Catholic education when the effect of educational level has been

TABLE 9.7

Score on Social Consciousness Scale*
(deviation from standard mean)

YEARS OF CATHOLIC EDUCATION	UNADJUSTED FOR EDUCATIONAL LEVEL	ADJUSTED FOR EDUCATIONAL LEVEL
1963		
$0_{(309)}$	− .19	− .09
$CCD_{(239)}$	− .03	− .04
1-11 years$_{(625)}$	− .05	− .01
12 years$_{(147)}$	.18	.06
13-15 years$_{(39)}$	1.07	.57
16-20 years$_{(21)}$	1.37	.62

eta = .12
beta = .06
Significance of Catholic education impact = n.s.

	1974	
$0_{(132)}$	− .67	− .52
$CCD_{(115)}$	− .14	− .08
1-11 years$_{(277)}$	.04	.07
12 years$_{(89)}$	.48	.31
13-15 years$_{(17)}$	1.12	.62
16-20 years$_{(14)}$	2.12	1.38

eta = .21
beta = .15
Significance of impact of Catholic education = .02.

*Attitudes toward Jews, blacks, social welfare, and charity toward others.

taken out.) Again we have a case in which the correlation between Catholic education and social consciousness is higher than the correlations used by the courts to justify rational integrated education.

In 1974 those with twelve years of Catholic education (presumably most of them with eight years of grammar school and four years of high school) were three-tenths of a standard deviation above the mean in social consciousness *even when their educational level was taken into account.* Those with thirteen to fifteen years were three-fifths of a standard deviation above the mean (approximately the same score as ten years previously), and those with more than sixteen years Catholic education were more than one standard deviation above the group mean in social consciousness. The principal difference between 1963 and 1974 is the expanded lead of those with twelve years of Catholic schools over the group mean at the latter period—.15 above those with no Catholic schooling in 1963 and .83 above them in 1974.

In summary, Catholic school enrollment continued to decline, mostly because of the unavailability of new schools and despite strong support for the schools in the Catholic population. Furthermore, Catholic schools have a considerable impact on adult behavior, particularly when compared with other forms of religious education and with measures of other educational outcomes. At the present time this impact is most notable among men, young people, and those from religious families; it is also felt strongly on organizational activities and personal devotion. It is less notable on ritual participation (such as mass attendance). Thus Catholic schools seem to be more important in a time of religious crisis for Catholicism, rather than less so. There is no evidence of Catholic schools being divisive. On the contrary, those who attend them seem to be more supportive of racial integration and have a higher level of "social consciousness" than those who do not attend Catholic schools. Far from being an obstacle to economic success, the Catholic schools seem to facilitate it.

There is nothing in the NORC data to support the mythology about Catholic education, and there is a good deal to support its efficacy. However, it is likely that the Catholic schools will not survive. Ironically, the Catholic hierarchy will do to the schools exactly what the nativists have wanted to do for a long time.

REFERENCES

1. Colin Greer, *The Great School Legend: A Revisionist Interpretation of American Public Education* (New York: Basic Books, 1972).
2. *The Official Catholic Directory* (New York: P.J. Kenedy & Sons).
3. For complete and detailed examinations of both studies see Andrew M. Greeley and Peter H. Rossi, *The Education of Catholic Americans* (Chicago: Aldine, 1966); and Andrew M. Greeley, William C. McCready, and Kathleen McCourt, *Catholic Schools in a Declining Church* (Kansas City: Sheed & Ward, 1976).
4. Greeley, McCready, and Court, *Catholic Schools in a Declining Church*, p. 255.
5. William C. McCready, "Faith of Our Fathers: A Study of the Process of Religious Socialization" (Ph.D. diss., University of Illinois, Chicago Circle Campus, 1972).
6. Greeley and Rossi, *Education of Catholic Americans*.
7. James S. Coleman et al., *Equality of Education Opportunity* (Washington, D.C.: U.S. Office of Education, 1966).
8. Marshal S. Smith, "The Basic Findings Reconsidered," in *On Equality of Educational Opportunity*, ed. Frederick Mosteller and Daniel P. Moynihan (New York: Random House, 1972), pp. 230–342.

9. F. Thomas Juster, ed., *Education, Income, and Human Behavior* (New York: McGraw-Hill, 1974), pp. 19–20.

10. Morris Rosenberg, "The Dissonant Religious Context and Emotional Disturbance," in *Religion, Culture, and Society,* ed. L. Schneider (New York: Wiley, 1964), pp. 549–559.

11. Juster, *Education, Income, and Human Behavior.*

12. Shirley Saldanha, "The Social Attitudes of Catholics and the Influence of Catholic Education: A Comparison between 1963–64 and 1974," multilith (Chicago: National Opinion Research Center, 1976).

CHAPTER

10

Family Structure

THERE IS a substantial social psychological literature investigating different patterns of family relationship among the three major denominational groupings, and the anthropological literature on peasant societies in Catholic European countries is fairly extensive. However, little attention has been paid by sociological researchers to whether differences in family behavior patterns persist among Protestants, Catholics, and Jews in the United States. Lenski determined, on the basis of a few questions administered to a small sample in one part of the country, that not only are Catholics more family oriented than Protestants, but that this family orientation impedes their social and economic progress.[1] As we have seen, there is little evidence that anything has impeded the economic progress of southern and eastern European Catholics; still, the question about the shape of their family relationships has not been answered in any depth.

Some aspects of family profile have been explored. We can see from Table 10.1 that Catholics are much less likely than either Protestants or Jews to be married before their twenty-first birthday; their divorce rate, like that of the Jews, is quite low compared with the British, Irish, and "other" Protestants. Catholic families are larger, and despite the conventional wisdom that Catholics believe a woman's place is in the home, the overwhelming majority of all the Catholic ethnic groups approve of working wives. Catholics are even more likely than Protestants to say they would support a qualified woman for the presidency.

Generally speaking, Catholics have more children than Protestants (Table 10.2), with the Irish and German Catholics having 2.6 and 2.7 children respectively. However, Polish and Italian family size is close

TABLE 10.1

Family Structure Variables for Religio-Ethnic Groups—Metropolitan Northern White
(percent)

ETHNIC GROUP	MARRIED BEFORE 21 YEARS OLD	DIVORCED	3 OR MORE CHILDREN	APPROVE WIFE'S WORKING	SUPPORT WOMAN FOR PRESIDENT
Protestants					
British	21	19	32	75	73
German	44	11	34	67	74
Scandinavian	47	9	41	59	72
Irish	61	28	45	64	73
"Other"	58	19	47	71	68
Catholics					
Irish	20	7	49	70	84
German	23	7	58	81	84
Polish	12	4	44	64	76
Slavic	13	9	46	77	85
Italian	24	11	47	61	79
Jews	40	5	24	77	94

to the national mean, and the Slavic Catholic family is about the same size as the Irish Protestant family. Thus, only the Irish and German Catholics appear to be notably above the mean in size of family. Furthermore, changing Catholic attitudes toward birth control (documented earlier) indicate that Catholic families of the future may be even smaller than Protestant families. It would appear that the larger Catholic family

TABLE 10.2

Number of Children
by Ethnic Group

Protestants	
British	2.1
German	2.1
Scandinavian	2.2
Irish	2.6
"Other"	2.3
Catholics	
Irish	2.6
German	2.7
Polish	2.3
Slavic	2.5
Italian	2.3
Jews	1.7

Mean = 2.4

size is limited to those women between thirty-five and forty-five; under thirty-five, Catholic married women may be having fewer children than Protestant women. Although there are not enough cases to look at the Irish and German family size by age on Table 10.3, it seems certain that there has been a considerable decline in the number of children produced in these two groups since the Vatican Council.

TABLE 10.3

*Number of Children and Age at Interview
for Protestants and Catholics*

AGE	PROTESTANT	CATHOLIC
18-24	0.4	0.4
25-29	1.5	1.3
30-34	2.4	1.8
35-39	3.0	3.2
40-44	3.2	3.4
45 or older	2.8	2.8
Number of cases	798	359

My colleagues and I at the Center for the Study of American Pluralism are convinced that family relationships and behavior patterns are extremely important if we wish to understand the transmission of ethnically linked behavior traits across generational lines. In the absence of research specific to this question, we are forced to piece evidence together from many different surveys. This technique enables us to chart a path through the obscurity of ethnic family structures, but necessarily raises almost as many questions as it answers. We are led to conclude that religious and ethnic family structures are much more complicated and intricate than had been thought.*

In his review of the family structure literature, Straus argues that much of the research indicates that two "reference axes" may be used to array patterns of interaction or personality. Relying on the work of many authors, he contends that the two principal axes in the family structure are power and support. He defines *power* as "actions which control, initiate, change, or modify the behavior of another member of the family." He defines *support* as "actions which establish, maintain, or restore, as an end in itself, a positive effective relationship with another family member."[2]

* Investigation of these questions is also hampered by the fact that in any given national sample, relatively small numbers of each of the individual religio-ethnic groups appear. Hence, explorations are even more chancy. The technique we generally use is to draw from the available anthropological and psychological literature a number of hypotheses about the various ethnic groups, and then see whether the survey data files available to us confirm such hypotheses.

Professor M. Kent Jennings, of the University of Michigan, made available to us a data set constructed from a study he did in 1965 of a national sample of American adolescents and their parents.[3] A number of questions were asked of both generations about the distribution of decision-making power in the family and about the psychological closeness of various members of the family to one another. Factor analysis produced two scales that correspond quite closely to Straus' support and power axes. The Jewish family (Table 10.4) was both very democratic (scoring thirty-six standardized points below the mean on power) and very affectionate and supportive (scoring twenty-two standardized points above the mean on support). This is a finding in keeping with the literature on the Jewish family. The Catholic family was somewhat higher on the power factor—that is, somewhat more centrally controlled —than the Protestant (seventeen standardized points), and slightly more supportive (seven points) than the Protestant. Thus the Jewish family is more democratic and more supportive than the other two families, whereas the Catholic family is more centralized and more supportive than the Protestant.

TABLE 10.4

*Protestant, Catholic, Jewish Scores on
Power and Support Scales*

	POWER	SUPPORT
Protestants(950)	02	−03
Catholics(316)	15	04
Jews(50)	−36	22

But if one looks at the various ethnic groups within Catholicism (Table 10.5), most of the variation from the mean on both power and support is accounted for by the Italian family structure, which is both strongly centralized and strongly supportive (fulfilling our prediction made from the available literature on the Italian family structure). The eastern European Catholic groups are also above the mean on both these factors, and the Catholic average on supportiveness is lowered because of the relatively low score of the Irish (minus nineteen standardized points). If one compares the British Protestant score and the Italian score on supportiveness, one can observe that the Italians are forty-six points higher. Eastern European Catholics are thirty-six points higher than Scandinavians on supportiveness. The eastern and southern European Catholics have very strongly supportive families, and the Italians, in addition, have strongly centralized families. There are striking differences between Catholic and non-Catholic family organization, espe-

TABLE 10.5

*Scores on Power and Support Scales for
American White Religio-Ethnic Groups*

ETHNIC GROUPS	POWER	SUPPORT
Protestants		
British(218)	00	−12
German(179)	−04	−10
Scandinavian(58)	−09	−24
Irish(64)	00	−21
"Other"(410)	07	− 2
Catholics		
Irish(44)	08	−19
German(54)	03	3
East European(35)	05	12
Italian(66)	19	34

cially when one compares the various ethnic groups within the two denominational groupings.

Another set of data taken from the NORC neighborhood study indicates that Catholics spend more time with their families (Table 10.6A). Regardless of their social class (as measured by the occupational prestige scale), Italians are the most likely to visit with parents, in-laws, and siblings every week. Indeed, more than half the Italians in the sample claim to visit those relatives every week, and, with one or two exceptions, all the Catholic groups are more likely to visit relatives more often than the Protestant groups.

It might be argued that patterns of family visiting are not so much the result of different family expectations and habits as the result of different geographical locations. Catholics, particularly Italian Catholics, are more likely than either Protestants or Jews to live in the same neighborhood or the same city as their parents and siblings. However, the findings presented in Table 10.6A do not change when one looks at the geographical distance from one's relatives (Table 10.6B). A third of the Italian Catholics whose parents live in different cities shall visit them every week (though in this case a "different city" may simply mean a ride across the Hudson River), as opposed to 12 percent of the English Protestants. Eighty-seven percent of the Italian Catholics who live in the same city but a different neighborhood visit their parents weekly, as opposed to 60 percent of the English Protestants and 62 percent of the Jews.

Family ties for American Catholics, then, are not only stronger

TABLE 10.6A

Ethnic Behavior by Social Class
(percent)

ETHNIC GROUPS BY RELIGION	N	LOW (DUNCAN OCCUPATIONAL PRESTIGE SCALE [1-5])					HIGH (DUNCAN OCCUPATIONAL PRESTIGE SCALE [6-10])				
		HIGH ON SOCIALIZING	VISIT WITH PARENTS EVERY WEEK	VISIT WITH IN-LAWS EVERY WEEK	VISIT WITH SIBLINGS EVERY WEEK	N	HIGH ON SOCIALIZING	VISIT WITH PARENTS EVERY WEEK	VISIT WITH IN-LAWS EVERY WEEK	VISIT WITH SIBLINGS EVERY WEEK	N
Catholics											
Italian	(178)	33	86	65	69	(76)	38	70	55	53	
Irish	(121)	39	57	51	55	(108)	45	40	47	42	
German	(99)	32	62	51	40	(50)	38	26	26	15	
French	(71)	20	62	65	48	(25)	30	52	64	37	
Polish	(52)	21	72	55	43	(19)	38	51	41	50	
Protestants											
English	(140)	34	43	41	32	(150)	44	32	28	20	
German	(99)	24	50	53	40	(95)	46	38	27	22	
Scandi-navian	(23)	22	47	30	26	(21)	59	27	32	17	
Jews	(49)	30	51	56	44	(113)	43	60	60	34	

SOURCE, tables 10.6A and B: Andrew M. Greeley, *Why Can't They Be Like Us* (New York: E.P.Dutton & Co., 1971), pp. 213, 214.

TABLE 10.6B

*Interaction with Relatives by Closeness
of Residence of Ethnic Groups by Religion*

ETHNIC GROUP BY RELIGION	INTERACTION WITH PARENTS WHO LIVE IN...		
	SAME NEIGHBORHOOD	SAME CITY, DIFFERENT NEIGHBORHOOD	DIFFERENT CITY
Catholics			
Italian	$100_{(66)}$	$87_{(62)}$	$33_{(46)}$
Irish	$100_{(22)}$	$70_{(45)}$	$18_{(69)}$
German	$100_{(12)}$	$76_{(41)}$	$15_{(54)}$
French	$95_{(22)}$	$77_{(22)}$	$26_{(31)}$
Polish	$100_{(6)}$	$86_{(20)}$	$26_{(13)}$
Protestants			
English	$96_{(35)}$	$60_{(43)}$	$12_{(104)}$
German	$100_{(16)}$	$74_{(36)}$	$21_{(81)}$
Scandinavian	$100_{(4)}$	$87_{(8)}$	$5_{(19)}$
Jews	?? $91_{(12)}$	$62_{(48)}$	$37_{(27)}$

for children and adolescents, as the data from Professor Jennings would indicate, they also persist into adult life. Catholics feel a greater obligation or need for, or a great delight in, interacting with their parents and siblings than do Protestants and Jews.

Differences in ethnic family behavior patterns are both subtle and blatant. Anyone who lives in a neighborhood with members of a different religio-ethnic group quickly becomes conscious that there are different expectations for appropriate family behavior in that group. On the other hand, assumptions about proper family obligations are so deeply embedded in the personality that it is hard to concede legitimacy to expectations and obligations that are different. "He's not really serious about visiting his mother every Sunday, is he?"

Lenski's assumption that close Catholic family ties would be a barrier to economic achievement is an excellent example of the difficulty of conceding "real" legitimacy to someone else's assumptions of what is appropriate in family life. Catholics, he observes, have a much closer and tighter family life, and *therefore* are less likely to be economically successful. Why the "therefore"? What is it about close family ties that impedes economic success? Why is a warm, centralized family an obstacle to economic achievement? The answer, I suspect, is that it is because Protestants do not behave that way and they are economically successful. One might observe that there may be other routes to economic success.

A third data set made available by Dr. Melvin Kohn of the National Institute of Mental Health (from data collected for him by NORC) confirms the general picture of a more active and more affectionate family life among Catholics (Tables 10.7A, 10.7B).[4] The Irish, Italian, and eastern European Catholic groups all report spending more time with their wives and children than do any of the non-Catholic groups. The German Catholics also report spending more time with their wives than any of the non-Catholic groups—though they are slightly less likely to spend time with their children than are the British-American Protestants.

There is more affection, support, and also more central control in Catholic families. Catholics are more likely to visit parents and adult siblings, and are also more likely to spend time with each other and

TABLE 10.7A

*Rank Order Among American White Ethnic
Groups on Time Spent with Children*

GROUP	RANK ORDER
Irish Catholics	14.069
Slavic Catholics	13.703
Italian Catholics	12.373
British Protestants	12.161
German Catholics	11.888
Jews	11.168
German Protestants	11.059
Scandinavian Protestants	10.243

SOURCE, tables 10.7A-D: Andrew M. Greeley, *That Most Distressful Nation: The Taming of the American Irish* (Chicago: Quadrangle Books, 1972), pp. 164, 165.

TABLE 10.7B

*Rank Order Among American White Ethnic
Groups on Time Spent with Wife*

GROUP	RANK ORDER
Italian Catholics	16.261
Slavic Catholics	14.324
Irish Catholics	13.971
German Catholics	13.760
Jews	13.535
German Protestants	12.623
British Protestants	12.560
Scandinavian Protestants	10.246

with their children than are Protestants. There is, however, some conflict between the Kohn-NIMH data and the Michigan data in the behavior of the Irish Catholic. The Michigan data indicated an Irish family which was low in support, while the NIMH data indicates an Irish family in which there is considerable time spent with wife and children. Of course, it is possible to spend time with family and not be supportive. The Irish Catholics are also the most likely in the NIMH data to hug, kiss, and scold their children. The other Catholic groups are spread out on the scale of affection and reproof for children. Italians are in the middle, eastern and southern Europeans toward the bottom, and German Catholics in the middle on affection and just behind the Irish on scolding. It may well be that displays of affection and reproof for a child are interpreted by the child in the whole context of the family relationship and may mean more in some groups than others. Still, Tables 10.7C and 10.7D raise some questions about Irish family life. One would expect the Irish father to be stern, reserved, and aloof, but it would

TABLE 10.7C

Rank Order Among American White Ethnic Groups on Amount of Hugging and Kissing of Children

GROUP	RANK ORDER
Irish Catholics	10.680
British Protestants	10.339
German Protestants	9.694
Italian Catholics	9.033
German Catholics	8.815
Slavic Catholics	8.547
Scandinavian Protestants	7.693
Jews	7.687

TABLE 10.7D

Rank Order Among American White Ethnic Groups on Amount of Child Scolding

GROUP	RANK ORDER
Irish Catholics	4.819
German Catholics	4.600
British Protestants	4.306
Italian Catholics	4.227
German Protestants	3.992
Scandinavian Protestants	3.878
Jewish	3.756
Slavic Catholics	3.450

appear that his relationship with his children is intense, combining a high level of affection with a high level of disapproval. Peculiar people, the Irish.

My colleague, William McCready, drew from the anthropological literature on the Italian, Irish, and Polish families a number of hypotheses to test against data from the national survey of women college graduates taken several years after their college graduation.[5] He predicted that the Italian women would report a warm and vigorous relationship with their mothers and would emphasize personal attractiveness and sex appeal in their own self-image. The Irish young women, he predicted, would be either at or below the national average on these measures. The Poles would be close to or beneath the national average on their relationships with their mothers but would emphasize, like the Italians, attractiveness and sex appeal. He finally suggested that the Italians would be little different from the national average in both achievement orientation and respect for the traditional role of women, while the Poles would be beneath the national average and emphasize domestic skills, and the Irish would strongly emphasize tradition but would not be nearly so positive on domestic skills as the Poles.

McCready's hypotheses were generally sustained. Italian young women have a warm affectionate relationship with their mothers, and strongly emphasized attractiveness. The Irish are traditional in their definition of the role of women, but apparently do not include within the tradition an emphasis on domestic skills. The Poles are extremely traditional, and, like the Italians, emphasize attractiveness and sex appeal, but they are substantially beneath the national average in reporting a warm and positive relationship with their mothers.

McCready's research has important implications for the study of ethnic diversity. If predictions made on the basis of the anthropological literature from the old country are sustained in survey analysis of young women during the 1960s in the United States, ethnic traits have remarkable durability. For our purposes, however, all that need be said is that there are not only differences between Catholics and Protestants in family attitudes and values, but these differences persist even among younger, college-educated women.

Do the different patterns of family relationships among the various Catholic ethnic groups indicate different expectations for children? One would expect that a strong centralized family of the sort the Italians, eastern Europeans, and Irish Catholics seem to have would lead these groups to expect obedience, courtesy, and honesty—the traditional marks of the "well-raised" child—from their offspring. However, it may also be that within certain cultures appropriate behavior for children may be defined differently, and that the honesty, politeness, and obedience supposedly inculcated by the strong family is more a British-

TABLE 10.8

Rank Order of American White Ethnic Groups on Values About Children Indicating Emphasis on Discipline and Control

GROUP	RANK ORDER
Scandinavian Protestants	−1.932
German Protestants	−1.263
Slavic Catholics	−0.906
Bristish Protestants	−0.047
Jews	0.604
Irish Catholics	0.699
Italian Catholics	0.783
German Catholics	2.551

SOURCE: Greeley, *That Most Distressful Nation*, p. 161.

NOTE: A high score indicates low emphasis on importance of child being honest, considerate, having good manners, being obedient, getting along, etc.

American tradition than a universal one. Table 10.8 indicates that this may well be the case. Discipline, control, obedience, and good manners are much more likely to be important for Protestants than for Catholics, though they are slightly more important for Slavic Catholics than for Protestants. The Irish, German, Italian Catholics have the lowest scores on the discipline control scale—even lower than the Jews. Catholics have very strongly controlled families, it would seem, but discipline and obedience for children does not seem nearly so important to them as to Protestants or Jews (with the exception of the eastern European Catholics).

Here there may be a hint of an explanation of why Lenski's assumption about authoritarian families impeding economic success is not sustained by the data. It may well be that in Catholic families, parents make the decisions (particularly the father) with much less consultation and democratic participation from children than would be the case in either Jewish or Protestant families; but it may also be that Irish and Italian Catholics (and Germans, too, apparently) are much less upset than are Protestant or Jewish families when parental injunctions are violated. One may have here a reflection of the differences in approach of systems of law, the Anglo-Saxon common law having a good deal more respect for obligation than does Roman ecclesiastical canon law. The former is much more empirical in deriving its obligations, but it expects them to be obeyed; the latter is much more a prioristic in its obligations, but gives wide latitude and exception, dispensations, and indults. Subject to much further research, it could be that a different style of democracy can be found in Irish and Italian families, in which more rules are made by parents with less consultation, but less concern is manifested about fulfilling the obligations of the rules. Certainly there are some impressions to be gained from the literary accounts of Irish and Italian families to suggest that this might be the case.*

Do these differences in family relationships, family obligations, family expectations have any impact on the personalities of those who have grown up in different familial interpersonal environments? Are they just interesting sidelights to American life, or do they have some critical effects on personality and behavior beyond the family environment? Both the NIMH and the Michigan data sets provide some important hints that family structures do indeed matter.

First of all (Tables 10.9A–10.9F), there are striking differences in the scores of various religio-ethnic groups on certain personality measures that Dr. Kohn and his colleagues at NIMH developed for their research. The Slavic, Italian, and German Catholic groups all score higher on a scale that measures "authoritarianism," though the Irish Catholics, who came from the most rigid and "cold" families, score rather low on this scale. On the other hand, on what is called the "moral liberalism," which could just as easily be called a "moral cynicism" scale ("If something works, it doesn't matter whether it's right or wrong"), the Irish are the most moralistic. Jews and the Italian and eastern European Catholics have the least moral hangups. The warmth and support of their family lives seems to provide both the Catholic and Jewish respondents with lower scores on "self-depreciation" than Protestants, and higher scores on "self-confidence." Catholics and Jews are much more likely to be "fatalistic" than Protestants, though the Italians do not seem particularly high on fatalism (despite the findings of other researchers). The Irish are the highest on "trust," a variable which does not seem to have any particular religious correlation (save that the Jews are lowest on it). How it is possible to be high on trust and high on fatalism, as the Irish are, is abundantly clear to anyone who was raised an Irish Catholic Christian. It is a culture which enables one to cry at baptisms and laugh at wakes. The world is an awful, evil, ugly, terrible place in which all kinds of terrible things will happen to you, but still, somehow, it will be all right in the end. The "great Gaels of Ireland" are truly, perhaps, the "men that God made mad," as G. K. Chesterton characterized them.[6]

---

* In my experiences as a parish priest working with mostly Irish but some Italian upper-middle-class families, I was forced to conclude that there were two kinds of rules, those that were made for the record and those that were serious. You were supposed to obey the latter and not get caught (at least not too often) in disobeying the former. I suspect that this form of adjustment to the generational conflict would horrify Protestants and Jews.

TABLE 10.9A*

Rank Order of American White Ethnic Groups on Personality Values That Indicate "Authoritarianism"†
(range = −4 to +4)

Items:
Young people should not be allowed to read books that confuse them.
In this complicated world the only way to know what to do is to rely on leaders and experts.
People who question the old and accepted ways of doing things usually just end up causing trouble.
There are two kinds of people in the world, the weak and the strong.
Prison is too good for sex criminals; they should be publicly whipped or worse.
The most important thing to teach children is absolute obedience to their parents.
No decent man can respect a woman who has had sex relations before marriage.

GROUP	RANK ORDER
Slavic Catholics	−1.535 (112)
Italian Catholics	−1.525 (130)
German Protestants	−0.205 (221)
German Catholics	−0.057 (122)
British Protestants	0.014 (1,290)
Irish Catholics	1.240 (102)
Scandinavian Protestants	1.590 (133)
Jews	3.349 (86)

SOURCE, tables 10.9A-F: Greeley, *That Most Distressful Nation*, pp. 147-152.

NOTE: A minus score indicates high "authoritarianism"; a plus score indicates low "authoritarianism."

*In Tables 10.9A-10.9F, the means are *standardized for social class*. All respondents are males.

†The scores in these tables are based on scales derived from a factor analysis of fifty-seven items. Only the items with a factor loading of more than .200 are listed.

TABLE 10.9B

Rank Order of American White Ethnic Groups on "Moral Liberalism" Scale

Items:
When you get right down to it, no one cares much what happens to you.
If something works, it doesn't matter whether it's right or wrong.
It's all right to get around the law as long as you don't actually break it.
It's wrong to do things differently from the way our forefathers did.
Once I've made up my mind I seldom change it.
You should obey your superiors whether or not you think they are right.
It's all right to do anything you want if you stay out of trouble.
It generally works out best to keep doing things the way they have been done before.
Do you believe that it's all right to do whatever the law allows, or are there some things that are wrong even if they are legal?

Jews	4.096
Italian Catholics	3.602
Slavic Catholics	3.189
German Catholics	− .413
Scandinavian Protestants	−1.450
British Protestants	−1.453
German Protestants	−1.660
Irish Catholics	−2.243

NOTE: A plus score indicates high "moral liberalism"; a minus score indicates high "moralism."

TABLE 10.9C

Rank Order of American White Ethnic Groups on Personality Values Indicating "Self-Depreciation" (range = −4 to +4)

Items:
I feel useless at times.
I wish I could be as happy as others seem to be.
At times I think I am no good at all.
There are very few things about which I'm absolutely certain.
If you don't watch out people will take advantage of you.
How often do you feel that there isn't much purpose to being alive?

GROUP	RANK ORDER
Scandinavian Protestants	−2.360
British Protestants	−0.548
German Protestants	−0.375
German Catholics	0.451
Irish Catholics	0.508
Slavic Catholics	2.020
Italian Catholics	2.516
Jews	3.519

NOTE: A minus score indicates high "self-depreciation"; a plus score indicates low "self-depreciation."

TABLE 10.9D

Rank Order of American White Ethnic Groups on Personality Values Which Indicate a Low Level of "Self-Confidence" (range = −4 to +4)

Items:
I generally have confidence that when I make plans I will be able to carry them out.
I take a positive attitude toward myself.
I feel that I'm a person of worth, at least on an equal plane with others.
I am able to do most things as well as other people can.
I become uneasy when things are not neat and orderly.
Once I've made up my mind I seldom change it.

GROUP	RANK ORDER
Slavic Catholics	1.205
German Protestants	0.765
British Protestants	0.498
Italian Catholics	0.479
Irish Catholics	0.161
Scandinavian Protestants	−0.183
German Catholics	−0.759
Jews	−0.901

NOTE: A plus score indicates low "self-confidence"; a minus score indicates high "self-confidence."

TABLE 10.9E

Rank Order of American White Ethnic Groups on Personality Values Which Indicate "Fatalism"
(range = −4 to +4)

Items:
To what extend would you say your are to blame for the problems you have—mostly, partly, hardly at all?
Do you feel that most of the things that happen to you are the results of your own decisions or of things over which you have no control?
When things go wrong for you, how often would you say it's your own fault?
How often do you feel that you are really enjoying yourself?
How often do you feel bored with everything?
How often do you feel guilty for having done something wrong?

GROUP	RANK ORDER
Irish Catholics	−1.971
Slavic Catholics	−1.378
Jews	−1.240
German Catholics	−0.810
Italian Catholics	−0.540
German Protestants	0.013
Scandinavian Protestants	0.720
British Protestants	1.345

NOTE: A minus score indicates high "fatalism"; a plus score indicates low "fatalism."

TABLE 10.9F

Rank Order of American White Ethnic Groups on Personality Values Which Indicate "Trust"
(range = −4 to +4)

Items:
It's all right to get around the law so long as you don't actually break it.
Human nature is really cooperative.
You should be able to obey your superiors whether or not you think they are right.
If you don't watch out, people will take advantage of you.
Do you think most people can be trusted?
How often do you feel that you can't tell what other people are likely to do, at times when it matters?

GROUP	RANK ORDER
Irish Catholics	2.506
Scandinavian Protestants	1.583
Slavic Catholics	1.481
German Protestants	0.767
German Catholics	0.757
Italian Catholics	0.502
British Protestants	0.242
Jews	−3.106

NOTE: A plus score indicates high "trust"; a minus score indicates low "trust."

Neither does there seem to be a particular religious pattern of the ranking of the groups on scales measuring "inner-direction" and "other-direction." Jews are high on inner-direction and low on other-direction; Italians are low on inner-direction and in the middle on other-direction; eastern Europeans are in the middle on both. The Irish, maintaining their inconsistency, manage to be high on both scales (Tables 10.10A, 10.10B). Insofar as the Kohn scales do indeed measure personality, one might suggest that it would appear that the warm and supportive Italian family life seems to produce adults who are in some ways authoritarian but in other ways morally liberal, not to say cynical. They have a moderately high level of self-esteem, a low level of trust, and a very low level of inner-direction. Such findings are not necessarily inconsistent with the picture we find in the sociological and anthropological literature, but a good deal more research needs to be done to confirm the present tentative findings and to sort out the mechanisms by which family structure produces personality styles.

TABLE 10.10A			TABLE 10.10B	
Rank Order of American White Ethnic Groups on Values for Self Which Indicate Emphasis on Self-Reliance (Inner-Direction)			*Rank Order of American White Ethnic Groups on Values for Self Indicating Emphasis on "Performance" (Other-Direction)*	
GROUP	RANK ORDER		GROUP	RANK ORDER
Jews	−1.594		Irish Catholics	−0.982
Irish Catholics	−1.158		German Protestants	−0.880
Slavic Catholics	−0.752		German Catholics	−0.749
German Protestants	0.008		Slavic Catholics	−0.324
Scandinavian Protestants	0.039		British Protestants	−0.252
German Catholics	0.191		Italian Catholics	0.323
British Protestants	0.498		Scandinavian Protestants	1.240
Italian Catholics	0.642		Jews	1.338

SOURCE, tables 10.10A and B: Greeley, *That Most Distressful Nation*, pp. 154, 155.

NOTE: A high score indicates low emphasis on interest, responsibility, self-reliance, common sense, importance of facts, coping with pressure.

NOTE: A high score indicates low emphasis on people, helpfulness, truthfulness, doing well, respect, and success.

The eastern European Catholic family, somewhat like the Italian in its structure, is also somewhat like it in the personalities which seem to emerge. The eastern Europeans seem higher, however, on both self-confidence and fatalism, and higher on trust and inner-direction than are the Italians. The Irish family, apparently very active in its relationships and both centralized and unsupportive, produces a personality that seems a bundle of contradictions—low on authoritarianism and moralism, high on fatalism and trust, and also high on both inner- and outer-direction. Catholic families, in other words, are different, and different personalities seem to develop from these families. The melting pot has not burned away all the differences in expectations and values that remain after immigration, but we have only the dimmest perception of how these differing values, styles, and personalities are transmitted.

It is impossible to relate the family structures we have described previously to the personality measures that have just been discussed. However, because Jennings' work is concerned with political behavior, it is possible to ask whether there is a link between family patterns and the different political behavior of young people in the different American religio-ethnic groups. Irish adolescents, as we might have expected from the description of the Irish political style in Chapter 5, tend to be high on all of the scales Jennings has developed (Table 10.11). Only on the cynicism and ideological scales are the Irish adolescents lower than the average, and this is consistent with the picture of a highly active political group. The Italians are beneath the average on virtually all the scores. Just as the Irish are both well informed and active politically and have a strong sense of political efficacy, the Italians have little sense of political efficacy and are neither active nor well informed. Scandinavians are rather like the Irish, but a good deal more social. Finally, the Jews are extremely well informed and quite active but not so likely to be trusting.

There are two questions we must ask. Does the shape of the family

TABLE 10.11

Political Values Among Four American Groups by Generation

POLITICAL VALUE	IRISH CHILDREN SCORE*	ITALIAN CHILDREN SCORE*	SCANDINAVIAN CHILDREN SCORE*	JEWISH CHILDREN SCORE*
Social trust	07	02	18	−08
Civic tolerance	07	−02	10	16
Political efficacy	33	−04	10	35
Ego strength	26	−43	43	11
Cosmopolitanism	25	−01	41	35
Political cynicism	−10	−18	−16	−06
Political activity	28	−11	10	25
Political knowledge	28	−17	19	65
Ideology	−12	01	−12	−11

*Scores represent standardized points deviating from mean for generation. 1.00 = 1 standard deviation. Scores are the percentage of standard deviation above or below the mean.

structure affect the ability of parents to transmit their own political
values to their children, and does the family structure itself affect the
political behavior of adolescent children?

A two-by-two model like the one developed by Straus is presented
in Figure 10.1. Straus suggests that the most effective transmission of
parental values will take place in the upper left quadrant, where both
power and support are the strongest. He does not suggest, although it
seems to follow logically, that the second most effective direct socialization
will take place in the upper right quadrant, where, if power is low, sup-

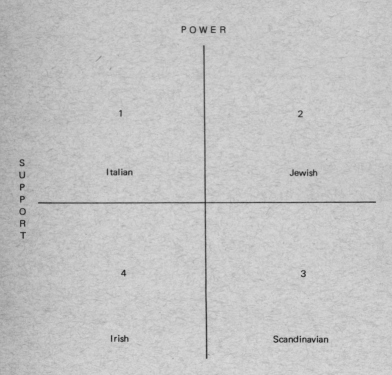

FIGURE 10.1

Power, support, and a hypothesized ranking of
"value transmission effectiveness" with the hypothesized
distribution of four American ethnic groups.

SOURCE: Andrew M. Greeley, "A Model for Ethnic Political Socialization," *The
American Journal of Political Science* 19 (May 1975), p. 195. (This and following
tables from this source are reprinted by permission of the publishers, Wayne State
University Press—© 1975.)

port is still high. Young people are not forced by power to imitate their parents, but are won by the relaxed and democratic styles of the family. They imitate their parents voluntarily. It would also suggest that the least effective direct socialization will occur in the lower left-hand corner, because in a situation where power is strongly concentrated in the family and support is weak, there will be a propensity to rebel against explicit parental values. Finally, in the lower right, where both support and power are low, socialization will be more effective than it will be in the rebellious lower left quadrant, but less effective than in the upper half of the diagram. The numbers we have placed in the quadrants indicate our expectations of the relative effectiveness of direct socialization.

Does family structure, as measured by power and support factors, predict the effectiveness of direct parent-child political socialization? Are the Italians the most effective in transmitting political values to their children, and the Irish the least effective? Table 10.12 indicates that the predicted ordering does exist in reality. The average intergenerational correlation on the nine political value scales is .215 for the Italians, .213 for the Jews, .146 for the Scandinavians, and .122 for the Irish Catholics. The Italians and the Jews are virtually tied for first place among the twelve religio-ethnic groups, and the Scandinavians and the Irish Catholics are at the bottom of the list, with their average intergenerational correlations being higher only than that of the German Protestants.

TABLE 10.12

*Rank Order of American Ethnic
Groups on Average Correlation
of Nine Political Value Scales
between Parents and Adolescents*

ETHNIC GROUP	SCORE
Italian Catholics	.215
Jews	.213
East European Catholics	.183
Irish Protestants	.172
German Catholics	.163
Spanish-speaking	.163
American Protestants	.160
Blacks	.152
British Protestants	.151
Scandinavian Protestants	.146
Irish Catholics	.122
German Protestants	.105

SOURCE: Greeley, "A Model for Ethnic Political Socialization," p. 198.

There are not only ethnic political subcultures, there are also ethnic family structures, some of which are much more favorably disposed than others to the direct transmission of political values.

The question arises: Might it be that in addition to facilitating or impeding the effectiveness of the family as a direct agent of political socialization, the structure of the family itself may be involved in the transmission of political values? For example, are Italian adolescents less likely to be politically involved not only because their parents are less likely to be involved, but also because there is something in the dynamic of Italian family life which leads—of itself and independent of parental values—to lower levels of political interest or concern? While there is little evidence in the general American society that family structure influences political values, perhaps within certain ethnic groups such influence is possible. We are clearly within a very speculative area, using measures designed for purposes other than those for which we use them here. Hence this phase of our analysis must be considered extremely tentative.

There is in the general population only one correlation between the two family structure scales and the nine political value scales we are using that is higher than .1, and that is a correlation of .12 between the power factor and the perception of sharp ideological differences between political parties.

But does family structure exercise such influence within the various ethnic subcollectivities? The Italian family is the best test case, because Italian political participation is low and the Italian family is strong on power and support, which might be thought to lend the family a self-sufficiency that would obviate the need for adolescents to involve themselves in other institutions.

Table 10.13 generally confirms this expectation. The stronger family support is in the Italian family, the weaker is trust, tolerance, cosmopolitanism, and political knowledge. Tolerance and political knowledge also correlate negatively with power in the Italian family, although political activity correlates positively with power. Its strong support mechanisms, then, incline offspring of the Italian family to be less concerned about the outside world (low on both cosmopolitan and political knowledge) and also more suspicious of the outside world (low on trust and tolerance). Furthermore, the strong concentration of power in the Italian family also correlates negatively with interest and positively with suspicion—though it is in those Italian families where power is most strongly concentrated that somewhat higher levels of political activity are likely to be found.

For the Jewish family one might expect just the opposite. Political participation and absence of suspicion might be heightened by the democratic style of the family and perhaps reinforced by its strong

TABLE 10.13

Correlations between Family Structure Scales and Political Values

POLITICAL VALUE	SCANDINAVIAN		IRISH		ITALIAN		JEWISH	
	SUPPORT	POWER	SUPPORT	POWER	SUPPORT	POWER	SUPPORT	POWER
Social trust	−.16	−	−	−.37	−.27	−	−	−
Tolerance	−	−	−	−	−.45	−.15	−	−.25
Efficacy	−	−	−	−	−	−	−	−
Ego strength	−	−	−	−.29	−	−	−	−.18
Cosmopolitanism	−	−	−	−	−.31	−	−	.26
Political cynicism	−	−	−	.24	−.21	−	−	−
Political activity	−	−	.20	−	−	.21	−	.20
Political knowledge	−.19	−	−.33	.16	−.18	−.27	−	−
Ideology	−	−	−	−	−	−.23	−	−

SOURCE: Greeley, "A Model for Ethnic Political Socialization," p. 209.
NOTE: Only correlations larger than .15 are shown.

internal cohesion and warmth. Indeed, it turns out to be the case that the more democratic the Jewish family (that is, the lower on the power scale), the more tolerant the adolescent and the more likely he is to be involved in political activity. Furthermore, ego strength (stubbornness), on which Jewish adolescents are much more likely than their parents to score high, also correlates negatively with power (and hence positively with democratic style). The more democratic the family, the stronger the ego strength of the Jewish adolescent. Power in the Jewish family correlates positively only with cosmopolitanism, and there are no correlations between political values and family support for the Jewish family.

It is somewhat harder to elaborate expectations for the Irish family. But one might predict that interest in politics for the Irish could result from strong family power constraints. The Irish young person is so busy calculating the implicit, unspoken conflicts over power in his family that he has a predisposition to be interested in the exercise of power wherever it occurs. However, it would only be in Irish families where there is some encouragement and warmth that children could work up enough courage to be actually involved in politics.

It is precisely in the high power-concentration Irish families that one is most likely to find higher scores of political knowledge—and also more cynicism, weakened ego strength, and weaker social trust. Furthermore, while the support level is low in the Irish family, it is in those families where there is more support than the average that the higher level of political activity is to be found, as well as a lower propensity to see sharp ideological cleavage between the parties.

Finally, one might ask whether the stronger "civic mindedness" of the Scandinavian adolescents might relate to the democratic structure of the Scandinavian family. The opposite seems to be the case. Civic trust (and political knowledge) relate negatively not to power but to support in the Scandinavian family. One finds higher levels of trust and of political knowledge in the Scandinavian families where affection seems to be the lowest. In three of the four cases, then, our tentative expectations about the relationship between family structure and political values were at least partially confirmed. Only with the Scandinavians could we find no support for our expectations.

It is interesting to note in Table 10.13 the different relationships between the family structures of various ethnic groups and political knowledge and political activity. For the Italians political activity correlates positively with a strong concentration of power in the family; for the Jews it correlates negatively. For the Irish there is a positive correlation between political activity and the support factor. Political knowledge correlates positively with power concentration for the Irish and negatively with power concentration for the Italians. It correlates negatively with support for both the Scandinavians and the Italians. The Irish who are the most likely to be politically knowledgeable come from power-concentrated families. The Scandinavians who are the most likely to be politically knowledgeable come from weak-support families. The Italians with high political knowledge come from families with weak power concentration and weak social support. The activist Italians come from high-power families, activist Jews from low-power families, and activist Irish from high-support families.

We conclude, then, that while the fit is not perfect and many uncertainties and obscurities remain, there is a relationship between family structure and political values in various ethnic collectivities. These relationships run one way in some ethnic groups and in the opposite direction in others.

There are ethnic subcultures which transmit different political values to the children born within these cultures. The ethnic subcultures have different family structures which affect the strength of the direct parent-child political socialization, with the family structures in some ethnic subcultures being much more successful at passing the explicit values of parents to the children. In addition, the family structure itself, independent of specific values of the parents, is also a socializing institution that transmits the ethnic culture. Italians, for example, are likely to be low political participators because their parents are low political participators, because the strength and warmth of the Italian family more effectively communicates low participation values across generational lines, and because the structure itself—independent of specific parental attitudes—produces lower levels of political interest and higher levels of political suspicion.

The Jews are high political participators in part because of higher social status, in part because Jewish parents are more likely to be high participators, in part because Jewish family structure facilitates the direct transmission of parents' participation values to their children, and finally because the democratic structure of the family itself, independent of specific parental values, creates a higher propensity for political activism.

Ethnic political structures, then, are handed on in part directly through children's imitation of their parents, in part both directly and indirectly because family structure increases the amount of imitation, and in part indirectly because family structure acts as an independent variable with an impact of its own for some ethnic groups on some political values.

One can elaborate a model of political socialization in which an adolescent's political behavior is shaped by three sets of influences: the values and behavior of his parents, the social class of his family as measured by parental education, and family structure as measured by power and support factors. One can then ask to what extent the differing family structures explain differing political behavior.

Jewish adolescents and Italian adolescents are rather different from each other in their scores on both political activity and civic tolerance, with the Jews in both cases having higher scores than the Italians. The difference in political activity is diminished by seventeen points by a control for parental education. None of the other factors in the model seem to have much impact. When parental education is taken into account, Jewish adolescents are still more likely than Italian adolescents to be politically active, though the difference has been narrowed somewhat.

However, on the subject of civic tolerance the model operates in another manner. Jewish adolescents are twenty-one points higher on civic tolerance than Italians. When parental values are taken into account, the difference diminishes to sixteen. When parental education is taken into account, the Italians actually become nine points more tolerant than the Jews; when the two family factors are taken into account, the Italian lead is increased to sixteen points. Social class, in other words, accounts for a change of twenty-five points, and the other variables in the socialization model account for sixteen additional points in the change (Tables 10.14 and 10.15).

Then there are some variables on which the model has almost no impact. The Irish are somewhat more likely than the Jews to score high on social trust (Table 10.12), but none of the controls introduced affect the difference very much. The slight propensity of Irish adolescents to be more trusting than Jews is not a function of education, parental

TABLE 10.14

Differences between Italian and Jewish Adolescents in Two Political Values
(Jewish "lead" in standardized points)

	POLITICAL ACTIVITY	CIVIC TOLERANCE
Raw differences	43	21
Net of parental value	41	16
Net of parental value and parental education	24	− 9
Net of parental value, education, and power factor	23	−11
Net of parental value, education, power factor, and support factor	24	−16

SOURCE: Greeley, "A Model for Ethnic Political Socialization," p. 202.

NOTE: Scores were obtained by multiple classification analysis with two categories (above and below the mean) for the value and factor scores, and three categories for education (grammar, high school, and college).

values, or family structure, and to the extent that it represents a meaningful difference at all, it must be attributed to some residual cultural difference not accounted for by our model.

The way we behave in our political life, then, is at least to some extent shaped both by what ethnic group we come from and by the family structures we have experienced in growing up, which themselves are typical of our ethnic group. Family structure does not explain everything (but then neither does religion and ethnicity), but there are differences in American society which are not caused primarily by social class differences. (Ethnicity is at least as strong a net predictor of scores on the Jennings political participation scale as is educational attainment.) Some of these differences do flow from different ethnic backgrounds and are shaped to some extent by the different family structures which are characteristic of these ethnic backgrounds. To pretend that this is

TABLE 10.15

Differences between Irish and Jewish Adolescents in Social Trust
(Irish "lead" in standardized points)

	SOCIAL TRUST
Raw differences	13
Net of parental value	10
Net of parental value and parental education	11
Net of parental value, education, and power factor	12
Net of parental value, education, power factor, and support factor	11

SOURCE: Greeley, "A Model for Ethnic Political Socialization," p. 202.

not the case or that such differences will eventually go away is a sign of commitment to an ideology or faith which must be held in the face of contrary empirical evidence.

All the data on which this chapter is based support the model of ethnic differentiation presented in Chapter 1. The family structure of American Catholics shares much in common with that of other religio-ethnic groups in the society. Nonetheless, there are some striking differences both between Catholics and non-Catholics and among the various Catholic subpopulations themselves. These differences have proved remarkably resilient in the face of pressures toward assimilation; they have not impeded the economic, educational, and occupational success of American Catholics. They do affect the way in which Catholics participate in American society, and they seem to persist across generational lines with remarkable tenacity. Proponents of restrictive immigration legislation after the end of World War I were concerned about whether the eastern and southern European immigrants would ever take on the values of "Americans" (by which they meant their own values, of course), and become good, successful, proper citizens of their new country.

What happened, apparently, was that the immigrants did accept the political consensus of their new homeland and did become economically and educationally successful—indeed, in a much shorter period of time than anyone could have imagined. But they did not give up all of their style and values and culture, and the differentials in those things would persist much more vigorously than anyone would have thought in the 1920s. The economic success took place sooner than was expected, and the cultural diversity persisted longer than was expected—and shows no sign of abating. The big surprise is that the persistence of cultural diversity did not interfere with the acquisition of educational and economic success. One has, then, neither a melting pot nor a mosaic, but rather a mosaic with permeable boundaries, or, in the words of the Mexican-American priest Father Elizandro, "a stew-pot" in which the groups remain somewhat distinct but nonetheless are powerfully influenced one by another and achieve some kind of rough parity with one another in the larger composite society.

What remains to be studied is the process by which some of the fundamental values of family structure are maintained while, at the same time, educational and economic success are achieved. The prediction was, as recently as the early 1960s, in Lenski's book, that either economic success would remain beyond the reach of Catholics, or they would give up their distinct ethnic values. In fact, the values are maintained and success still achieved. Do the ethnics compartmentalize their lives, or is it possible that the values of family and neighborhood (to be discussed in Chapter 11) may actually reinforce their economic achieve-

ment? My colleagues and I at the Center for the Study of American
Pluralism hope to pursue this subject in greater detail in the years
ahead.

So the Catholic ethnics are still different in the fundamental matter
of family structure, and they are likely to continue to be different. There
is, therefore, bound to be some tension between Catholics and others,
because family structure is so primordial that it rarely becomes a matter
for explicit discussion. Just as the different Catholic political participation styles that we discussed in Chapter 5 subliminally irritate many
non-Catholics, so is there every reason to assume that differential
Catholic family values also subliminally irritate non-Catholics—and, of
course, vice versa. Indeed, if one were looking for an explanation for the
stereotyped conventional wisdom which this book is endeavoring to
demolish, it might well be that much of the origin of subliminal anti-
Catholicism is based on the irritation of differential Catholic values and
styles in politics, family life, and neighborhood community. Yet it is not
altogether clear that America's cultural elite concede to Catholics the
right to be different in these matters. Present evidence indicates that
Catholics will not succumb to the pressures of conformity.

REFERENCES

1. Gerhard Lenski, *The Religious Factor*, rev. ed. (Garden City, N.Y.: Doubleday Anchor, 1963), p. 345.
2. Murray A. Straus, "Power and Support Structure of the Family in Relation
to Socialization," *Journal of Marriage and the Family* (August, 1964): 318–326.
3. M. Kent Jennings, *The Student-Parent Socialization Study* (Ann Arbor:
University of Michigan Inter-University Consortium for Political Research, 1971).
4. Melvin L. Kohn, *Class and Conformity: A Study in Values* (Homewood,
Ill.: The Dorsey Press, 1969).
5. William C. McCready, "The Persistence of Ethnic Variation in American
Families" in Andrew M. Greeley, *Ethnicity in the United States: A Preliminary
Reconnaissance* (New York: John Wiley & Sons, 1974), chap. 7.
6. G. K. Chesterton, "The Ballad of the White Horse," in *The Collected Poems
of G. K. Chesterton* (New York: Dodd, Mead & Co., 1932), pp. 205–299.

CHAPTER
11

The Neighborhood

AS WE HAVE SEEN in previous chapters, Catholics place a heavy emphasis on the values of family, local community, particularized political contact, dense interaction networks, nonideological particularism in their organizational style, and informal, casual, and indirect methods of doing things, rather than formal, bureaucratic, direct techniques. Catholics, it would seem, have some very different fundamental assumptions about the nature of human nature and the nature of society, assumptions which are not accepted by and often offensive to those who believe in a rationalized, formalized, bureaucratized approach to social organization and decision making.

Many Americans who are convinced of the need for social change and reform do not think it necessary to take into account the impact of the plans and programs of their rationalized decision-making processes on local communities which have been selected to be the targets of such programs. When the people who live in those communities react and resist, they are written off as bigots, racists, or relics of a bygone tribal age.*

There is no better example of the confrontation between the two social ethics than the neighborhood. In terms of the rational, formal ethic, the neighborhood is merely a place where one happens to live. It is a location an individual chooses to move into and from which he can

---

This chapter is a revised and expanded version of Chapter 8, "The Experience of the Neighborhood," from *The Communal Catholic: A Personal Manifesto*, by Andrew M. Greeley. Copyright © 1976 by The Seabury Press, Inc. Reprinted by permission.
* For a classic description of the two mentalities, see Mario Cuomo, *Forest Hills Diary: The Crisis of Low-Income Housing* (New York: Random House, 1974).

choose to move as a result of rational, individual, self-controlled decision-making processes. The "modern" mind simply cannot understand why a neighborhood should be viewed in any other light. Attempts to turn the neighborhood into the urban equivalent of the peasant village seem regressive and romantic. A neighborhood is wiped out by a social reform plan? So what?

But for those who are committed—however implicitly—to the ethic described in Chapter 10, neighborhoods are profoundly important. The individual family is not an isolated unit acting out of its own enlightened self-interest and making rational decisions in response to that self-interest. The individual family is part of a dense, complex, intricate, restrictive, yet supportive network of human relationships. For such a mentality, destruction of a neighborhood is a disaster.

For all its importance in American cities, the neighborhood has been the object of relatively little research in recent years. The early "Chicago school" of sociology talked a lot about neighborhoods and even observed some, and the work of Gerald Suttles (to be discussed later) represents a continuation of this approach; but Suttles is almost alone among contemporary sociologists in his concern for neighborhoods. The comparative study of neighborhoods based on survey data has been undertaken only once, in the NORC Integrated Neighborhood Survey.[1] Neighborhoods are obstacles, barriers, undigested lumps out of the tribal past. They are gradually being urban-renewaled out of existence, and they are not replaced in the new suburbs.* Hence it is impossible to write a chapter on the neighborhoods with the sort of massive survey evidence on which most of the other chapters have relied; we shall instead refer to historical and impressionistic evidence. Anyone who seeks to understand American Catholics would make a grave mistake in underestimating the importance of the neighborhood in the life of the Catholic collectivity. The new National Conference on Ethnic and Neighborhood Affairs did not casually add the term "neighborhood" to its title. As was evident to anyone who attended its founding session in the autumn of 1975, "neighborhood" was the decisive word.

We do have some empirical evidence, however. About half the Catholics in the country live in a neighborhood which is one-half or more Catholic—a proportion which has not changed since 1963. Twenty percent of American Catholics live in a neighborhood which is almost all Catholic. One-third of American Catholics live in a neighborhood in which more than half their neighbors are of the same ethnic group; one-third also live in a neighborhood where there are blacks—an

---

* The Catholic liberal conventional wisdom maintains this with vigor. Like most other Catholic liberal assertions, this one is totally devoid of any empirical evidence. Indeed, even impressionistic evidence runs to the contrary. Suburbs that are heavily Catholic do indeed become neighborhoods.

extraordinarily high proportion, given the alleged racism of Catholic ethnics. One-quarter of American ethnics live in or close to the neighborhood in which they grew up, and half the Catholic adults in the country grew up in a neighborhood which was more than half Catholic (the same proportion as ten years ago). Half also think that the neighborhood in which they live now is a very good one to live in. Catholics are also much more likely to live in the same neighborhoods as their parents and in-laws; one-third of the eastern and southern European Catholics live in the same neighborhood as their parents, as opposed to less than one-fifth of the rest of the population. About one-quarter report they also live in the same neighborhood as their in-laws.

Sociologists have a hard time defining a neighborhood. Herbert Gans uses the cultural approach and calls it "the urban village."[2] From the social psychological perspective, Gerald Suttles maintains that the neighborhood by definition is a place to be defended; it is that segment of the checkerboard of the city where one feels safe.[3] Demographers try to define it in socioeconomic terms; the U.S. Census thinks of the neighborhood as a census tract. Some researchers try to define a neighborhood empirically as that area which most of the people within it consider to be a neighborhood.

Unfortunately, subjective definitions of "neighborhood" gives rise to several different opinions of what the boundaries of the neighborhood are. Furthermore, the neighborhood exists on several different levels. The block on which one's neighbors live (across the alley in many of the "old" neighborhoods, across the street in many of the "new" ones), the one or two blocks surrounding one's own where it is considered "safe" for younger children to play, and the "broader community" from which the neighborhood may get its name and wider identification are all proper uses of the term "neighborhood." In the south side of Chicago, for example, the neighborhood may be for all practical purposes coterminous with the parish. Thus, young people asked where they are from would not say "Brainard" or "Englewood" or "Beverly" but "St. Killians," "St. Leo," or "St. Barnabas." Even Protestants seem to get into the habit of identifying their neighborhoods by the name of the Catholic parish, not so much out of ecumenical motives but simply, one supposes, for clarity of discourse. Some neighborhoods are large enough to have several parishes, and in "old" neighborhoods with national parishes, many shared the same territory.

So the term "neighborhood" is amorphous. The people who use it know what they mean and know in which of its many senses they are using it. The social scientist who hears them must be careful to understand in which sense the word is being used. Yet because the concept is ambiguous does not mean that the reality does not exist, as some social scientists have argued. It means merely that the reality is multilayered.

Human language is not designed to provide precision for sociological analysis. The messiness of language, however, corresponds better with the messiness of social reality than does the precise terminology of social science.

It is not clear to me whether cities other than those in America have neighborhoods, or even whether all American cities have them.* Insofar as I can tell, the boroughs of London, the arondissements of Paris, the districts of Rome, Vienna, and Berlin are not neighborhoods in the sense the word has acquired in the south side of Chicago, in Brooklyn, Boston, Detroit, St. Louis, Milwaukee, and other cities of the northeast and north central regions of the United States. However, James Duran describes the creation of enclaves in a city in Kenya in which tribes become ethnic groups.[4] Duran's ethnic enclaves sound suspiciously like neighborhoods. Perhaps "foreign" immigration and diverse tribes coming together in a city are analogous and is what is required for the "neighborhood," in the American sense, to come into being.

In its origins the neighborhood seems to have been the port of entry for immigrants. Neighborhoods may well have existed in Paris and London in the nineteenth century, but the country folk flocking to those two cities were, after all, English and French; they never became ethnic groups as did the immigrants to American cities and, it would seem, to Duran's Kenyan city. London would make an interesting case study. In addition to the official map provided by the boroughs, there surely is in modern London an unofficial map marking out the boundaries of the various ethnic enclaves in that multiracial, multiethnic, polyglot city. I have the impression that there are indeed neighborhoods in London, though perhaps they are not as clearly defined or as important for self-identity as their counterparts in the United States.†

I would suggest tentatively that the neighborhood as it is known in the northeast and north central parts of the United States is the result of the immigrants' experience. Unfortunately, we have little in the way of history of the development of the neighborhood or parish out of the immigrant experience. As Jay Dolan remarks, "For many years American Catholic historiography has focused on episcopal biographies and the internal controversies of the Church in the late nineteenth century. As a result there has been little writing done on the urban dimension of the American Catholic Church."[5]

* I have heard it argued that Los Angeles is devoid of neighborhoods. Such may well be true of the vast tract developments of the San Rafael and San Fernando Valleys; it may also be true, for all I know, of Orange County; it is surely not true of the city of Los Angeles itself.
† I am not suggesting that neighborhood is or was important to the self-definition of all its citizens. It is important for many and very important for a by no means small number.

One might add that the historiography of which Dolan writes is relatively unconcerned with the social dimensions of history and is out of touch with the mainstream of recent American historical thought. It is no discredit to Thomas McAvoy or John Tracy Ellis to say that the work they and their students have done represents a much earlier phase of American historiography, a phase in which facts and events are narrated with little attention paid to social or economic context. General theoretical models like "Americanization" or "the influence of the frontier" may be added to the narrative descriptions of the institutional histories and biographies, but they scarcely flow from the data or admit testing or further research. In fact, as Dolan points out, it is not the frontier but the city, not Americanization but cultural pluralism, that is the distinctive phenomenon of the American Catholic experience. In his study of the development of the Catholic neighborhoods and parishes in New York City from early to middle nineteenth century, Dolan establishes conclusively that the emergence of the national parish (there were none in New York in 1815; by 1865 they were one-third of the parishes in the diocese) was the unique adaptation of Catholicism to the pluralism of American cities.

Dolan writes:

In studying American Catholicism from an urban perspective the similar characteristics of different urban churches are strikingly exhibited. The unity of American Catholicism is thus more readily perceived and when combined with the cultural pluralism of its people, the genius of American Catholicism becomes more evident. Briefly stated, this genius may be described as unity in the midst of cultural pluralism. The unity is based on a common faith, a common church and a common country, and the unifying effect of the urban environment should not be underestimated. In each urban church there were common problems, similar national groupings, and similar patterns of organization together with similar educational and benevolent apostolates. Such similarities naturally linked the church in Chicago with the church in New York and in other large cities and this communality helped to shape the self-identity of American Catholicism. The identity began to take shape during the middle decades of the nineteenth century when urbanization in America proceeded at a rate faster than ever before.[6]

It was Archbishop John Hughes who recognized that nationalism was "an exceedingly tender and delicate topic." Dolan comments:

In encouraging the organization of national parishes in the city and the diocese, he lessened the possibilities of conflict and recognized the inherent differences between ethnic groups. During the middle decades of the century the national parish became the church's adaptation to the variety of people living in the city and as the immigrants continued to swell the urban population later in the century the national parish became even more necessary. When the church ignored the ethnicity of its people and failed to recruit ethnic priests and to establish national parishes, then conflict inevitably arose.[7]

But despite the national parish and the potential for disunity it involved, there were still the unifying factors of commitment to American society and its precious gift of freedom. Dolan notes:

Appealing to their adopted patriotism, Catholic spokesmen quickly acquired the style of a Fourth of July orator. The public exhibitions of parochial school students encouraged patriotic poems and songs; and the two great principles which guided the New York Catholic paper, Freeman's Journal, were to be truly Catholic and truly American.[8]

Characteristic of the American national parish was the organizational control exercised by the chancery office. Dolan implies that the reason why American bishops seemed to have so much more power vis-à-vis their priests than did bishops in other parts of the world is that a highly centralized diocesan administration was a sine qua non for holding together the diversity of their pluralistic dioceses.

Dolan's study is careful and precise. He shows the influence of omnibus lines and street railroads in the spread of the parish system up Manhattan Island as the immigrants poured into the city. By detailed examination of parish records, he draws an accurate demographic picture of the spread of the church in New York City. Unfortunately, we have very few such studies of the creation of immigrant parishes and neighborhoods, and of how church leaders tentatively and experimentally developed a response to the pluralism which had been dumped into their laps.

Nelson Callahan has made a detailed study of national parishes in Cleveland. He summarizes his research:

One hundred years ago, in the midst of their own massive immigration to this country, it was the Germans who first seemed to have fixed clearly in their own goals the insight on the connection between language, culture, custom, and religion. They said that if any of these four components was removed from the lives of the immigrants, the other three would disappear within one generation. Hence, they fought for and indeed achieved the German national parish, an anachronism today but once the model for all the other non-English-speaking Catholic ethnic groups that have come to this country since 1848. If one never was a member of such a parish (they actually were always referred to as "congregations" by our early bishops), one might have little knowledge of the dynamics that occurred in each group of immigrants as each sought to continue to transfer the message and person of Jesus while at the same time they experienced the trauma of immigration. What were the German parishes like? The answer was the same for each parish.

1. They had singing, sermons, and paraliturgical services in the German vernacular.
2. The people of these parishes were welded together in real communities; the parishes had no special territory, as did the English-speaking ones (which the Germans always referred to as the "Irish" parishes—never "American," for they seemed to know intuitively that acculturation had not taken place there either). The people were united with one

another by a covenant of worship which was acceptable to all. If a member did not like this covenant, he was free to join the territorial parish which was usually close by. In the first two generations, few people did this.

3. Decisions of these congregations that concerned the general life of the whole group were arrived at through a semidemocratic process which was peculiar to the United States. Seldom were the first pastors of these congregations as absolute in their parishes as were their sacerdotal brothers in the territorial parishes.

4. These parishes, almost without exception, began and supported parish schools in which religion was always taught in the language of the non-English-speaking homeland.

5. These parishes usually had pastors of the same ethnic background as that of the people; if no such pastor could be had, the pastor who did serve the congregation had to be one with his people in culture, language, and custom.

One would have to say that this system worked very well, since the fact is that it does seem to have preserved the faith of the immigrants. One can only speculate about whether or not it should have been continued. But there are two facts that seem to obtain true in every one of these national parishes (at least those in Cleveland). They seem to me to be interrelated.

1. Less than 5 percent of the immigrants or their children married outside of their nationality and less than 2 percent of them entered mixed marriages. I have, however, no data on what percent of the immigrants or their children married outside of the church, since my research covers only church records.

2. During the first generation of the life of any of these parishes, regardless of their nationality, the vast majority of them made no converts at all. I will cite one example: Our Lady of Lourdes Slovak congregation, which still is a viable parish in Cleveland, was founded in 1883. The parish built a monumental Gothic church, a large school, convent, and rectory; it was staffed by zealous priests who seem to have been exemplary men; it numbered over 2,000 families by 1900 and yet the first convert baptized in the parish was in 1916.

The same overall pattern seems to be true in all of the national parishes during the first generation of their existence. Now what is one to make of all of this? I would suggest that perhaps the people of the national parish may have felt that their parish existed to preserve their faith, language, culture, and custom values. At the same time, they felt that these values had no genuine, transferable force outside of the group. . . . I would suggest that those parishes had a genuine hesitancy about the stability of the marriages contracted by any of their members with the members of any other similar, but divergent, Catholic group, as well as a radical fear about the stability of the marriage of any of its members with anyone from another divergent and non-Catholic group. To marry inside the group was to avoid conflict in culture, to preserve a cultural sameness in marriage expectations, and to facilitate the ability to give true matrimonial consent.[9]

In the absence of research like Dolan's and Callahan's, we shall have to make do with that curious combination of anecdote and ideology which passes for Catholic history. The tentative description of the genesis of neighborhoods contained in the subsequent pages of this paper is at best a speculative model to be tested against the further research which will be done in the years to come.

Almost as soon as they arrived, the immigrants set about to organize themselves in a rough fashion to cope with the problems of their new environment. In many cases such organization was facilitated by the fact that the immigrants had come from the same place in the old country and brought not merely a culture but the skeleton of a social structure with them. The priest or the pastor came very early; so, too, I suspect, did the tavern owner and the shopkeeper; not too long after, the undertaker and the political leader arrived. The earlier groups did not bring professionals with them, save perhaps for an occasional doctor. Yet it would not take long for a young man of seventeen, if he was bright and ambitious, to find his way to medical school or law school and begin professional practice in the neighborhood. The teaching nuns, perhaps from the old country or, as in some cases, founded in the United States to recruit people from the old country, would begin the parochial school. Young women from the community would go to "normal" school and take up jobs teaching in the public schools, if not in their own neighborhood, then nearby.

The speed with which the community developed its own services must have varied from group to group and have been closely connected with the speed with which the group's members could acquire the English language.* One can go today into one of the older ethnic neighborhoods of a city and see the social structure of the immigrant neighborhood. For example, on South Lowe in Chicago (Mayor Daley's street, which ought to be preserved as a national monument) one can find within one block the police station, the fire station, the Catholic church, the politician's house (in this case a very important man), the undertaker's, the corner grocery store, and the tavern. In newer neighborhoods the tavern may be replaced by the country club and the corner grocery by the supermarket, but the other institutions have survived.

There are descriptions in Oscar Handlin's work[10] and others of the incredible poverty and degradation experienced by the early waves of Irish famine immigrants to Boston and New York City. The life of the first immigrants was not easy. My maternal grandfather died working in the sewers in his middle forties; his wife was dead of pneumonia

---

* Many of earliest Irish immigrants were Irish speakers. Some, it now appears, did not even speak English, while others spoke it as their second language. Still, even some English speakers within the group gave the Irish an advantage over continental immigrants. Surely by 1870 all the Irish immigrants must have been able to speak English at least as a second language. Recent linguistic research indicates that the so-called "Brooklyn" accent is a descendant of English spoken by those whose first language was Irish.

within a year, leaving the oldest girl, still in her early twenties, to raise the rest of the family. Finley Peter Dunne, in some of his early writings, tells what life was like for "little Tim Clancy," one of the early immigrants to Chicago's Bridgeport neighborhood:

He wor-ruks out in th' mills, tin hours a day, runnin' a wheelbarrow loaded with cindhers. He lives down beyant. Wan side iv his house is up again a brewery, an' th' other touches elbows with Twinty-Percint Murphy's flats. A few years back they found out that he didn't own on'y the front half iv th' lot, an' he can set on his back stoop an' put his feet over th' fince now. He can, faith. Whin he's indures, he breathes up th' chimbly; an' he has a wife an' eight kids. He dhraws wan twinty-five a day—whin he wurruks.

Marriage and family life in Bridgeport was not easy, either:

People that can't afford it always have marrid an' always will. 'Tis on'y th' rich that don't. They niver did. That's wan reason why they're rich, too. But whin a young man is so poor that he can't afford to keep a dog an' has no more prospects thin a sound-money dimmycratic newspaper [supporting William Jennings Bryan], he finds a girl who's got less an' proposes to her an' they're marrid at th' expinse iv th' grocers iv the neighborhood an' they live unhappy iver after, bringin' up a large fam'ly to go an' do likewise.

But there was also fun at church carnivals and, of course, on St. Patrick's Day:

Twas a g-grand fair. They had Roddy's Hibernyun band playin' on th' corner an' th' basemint iv th' church was packed. In th' baa-ack they had a shootin' gall'ry where ye got five shots f'r tin cints. Hogan, th' milkman, was shootin' whin I wint in an' iverybody was out iv th' gall'ry. He missed eight shots an' thin he thrun two lumps iv coal at th' ta-arget an' made two bull's-eyes. He is a Tipp'rary man an' th' raison he's over here is he hit a plisman with a rock at twinty ya-ards—without sights.

I'd no more thin inthered th' fair thin who should come up but Malachi Dorsey's little girl, Dalia. "Good avnin' " she says. "Won't ye take a chanst?" she says. . . . Whin I come away I stood to win a doll, a rockin' chair, a picture iv th' pope done by Mary Ann O'Donoghue, a deck iv ca-ards an' a tidy [bear].*

Th' booths was something iligant. Mrs. Dorsey had th' first wan where she sold mottoes an' babies' clothes. Next to hers was the ice crame lay-out, with th' Widow Lonergan in cha-arge. . . .

Acrost th' hall was th' table f'r church articles, where ye cud get "Keys in Hevin" and "St. Thomas a Kempises" an' ros'ries. It done a poor business, they tell me, an' Miss Dolan was that sore at th' eyesther shtew thrade done be Mrs. Cassidy next dure that she come near soakin' her with th' "Life iv St. Rose iv Lima." 'Twas tur-r-rible. . . .

Displaines street . . . south to Harr'son, wist to Bloo I'land avnoo, south-wist to Twilfth, where th' procission'll counthermarch befure th' Jesuit

---

* These quotations from Dunne's Mr. Dooley are from a superb article by Charles F. Fanning, Jr., "Mr. Dooley's Bridgeport Chronicle," *Chicago History* 2 (Spring, 1972): 47–57. Anyone interested in a portrait of immigrant life by a shrewd observer should read the Fanning article. I hope it develops into a book.

Church an' be reviewed be his grace th' archbishop, be th' clargy an' th' mayor an' th' board iv aldhermin.

Attintion! Carry ar-rms. Where's th' band? Officer Mulcahy, go over to Dochney's an' chop that band away fr'm th' bar. Hol' on there, Casey don't back that big saw horse again me. Ma, look at da-da in Gavin's hack. Ar-re ye ready? Play up th' wearin' iv th' green, ye baloon-headed Dutchmin. Hannigan, go an' get th' polis to intherfere—th' Sons iv Saint Patrick an' th' Ancient Order's come together. Glory be, me saddle's sliipin'. Ar-re ye ready? For-wa-ard march!

However rough life may have been in Bridgeport when the young Finley Peter Dunne and the saloonkeeper Dooley observed it, it was much better organized and much less degrading than the conditions in New York and Boston twenty years earlier. Irish immigrants were still coming in in the 1880s, though not at such a rapid rate as in previous years. But now there was a community organized to receive them and to orient them toward the new society. One of the fascinating and as yet unresearched historical questions is how long it took each of the groups to establish a "receiving" community which could protect the newest immigrants from abject poverty and exploitation when they arrived on American shores. Certainly the letters in Thomas and Znaniecki's *Polish Peasant in Europe and America*[11] would lead one to believe that such a rough but serviceable receiving community had been built by the Poles in Chicago as early as 1910.

There is no point in glamorizing the immigrant neighborhood. Life was difficult, but it was much less difficult for many of the immigrants than it had been in the country they left behind. Few seemed disposed to go back, in any case.

The neighborhood soon became the source of power and strength. It was the immigrant's political base—tenuous and very rough, at that:

Whin Andy Duggan r-run f'r aldherman against Schwartzmeister, th' big Dutchman,—I was precinct captain then, Jawn,—there was an iliction f'r ye. Twas on our precinct they relied to ilict Duggan; f'r th' Dutch was sthrong down be th' thrack, an' Schwartzmeister had a band out playin' "Th' Watch on th' Rhine." Well, sir, we opened th' polls at six o'clock, an' there was tin Schwartzmeister men there to protect his intherests. At sivin o'clock there was only three, an' wan iv thim was goin' up th' sthreet with Hinnissy kickin' at him. At eight o'clock, be dad, there was on' wan' an' he was sittin' on th' roof iv Gavin's blacksmith shop, an' th' la-ads was thryin' to borrow a laddher fr'm th' injine-house f'r to get at him. . . . We cast twinty-wan hundhred votes 'r Duggan, an' they was on'y five hundhred votes in th' precinct. We'd cast more, but th' tickets give out. . . .

The neighborhood also became an economic base. Craftsmen, storekeepers, undertakers, schoolteachers, saloonkeepers like Dooley, contractors, deliverymen, and the occasional doctor, lawyer, or dentist began to amass moderate and eventually substantial incomes by serving the needs of the immigrant community. You tended to do business with "your own kind."

And of course there was the parish, protecting the religious faith of the people even on those occasions when the faithful didn't want protection. It also not infrequently provided for the people's material welfare and promoted their political advance. Even the normally skeptical Mr. Dooley had nothing but praise and admiration for his parish priest "Father Kelly."* Georg Mann observes:

If there is a hero in the Dooley columns it is Father Kelly, the parish priest, a warmly human and sensible man. When Molly Donahue, convent educated in the emerging upwardly mobile immigrant tradition, creates a parish scandal by riding a bicycle in a divided skirt, Father Kelly calmed the commotion by telling Molly that he hadn't been aware before that she had bowlegs, and the divided skirt disappeared. Capable of striking an obnoxious parishoner [sic], and repenting for his outburst of temper afterward, Father Kelly believes that charity is more important than the strict observance of the rites of the Church. And he even allows Dooley to persuade him to take a drink during Lent.[12]

Charles Shanabruch describes Father Dorney of St. Gabriel's (in the "back-of-the-yards" district in Chicago) as the "absolute and un-questioned lord and master of the entire district."[13] An autocrat Dorney may have been, but he was also a fierce defender of his people and of the trade union as a means of protecting the rights of his people.

We have practically nothing in the way of studies of the style and methods of the immigrant pastors like Father Dorney. An important exception is Daniel Buczek's *Immigrant Pastor*,[14] a study of the life of Monsignor Lucyan Bojnowski, a Polish pastor in New Britain, Connecticut. The book is a fundamental resource for any study on the praxis of the neighborhood. A brilliant, sophisticated aristocrat, Bojnowski was by conviction as well as temperament a pluralist. Unlike his fellow Polish American Francis Hodur, he did not believe that the rights of Polish Americans could be defended against the Irish-dominated church by schism. Nor did he agree with his other contemporary, Waclaw Kruszka, that constant confrontation with the Irish hierarchy was essential. Bojnowski was capable of telling off bishops either by telephone or in person, and of appealing over their heads to the apostolic delegate. The sight of him, canon law book under one arm and commentary on the other, charging into the New Haven Chancery Office, was apparently a common one. Bojnowski mixed confrontation with civility, respect, and a punctilious regard for protocol. Some of his Irish bishops must have wished on occasion that he had gone into schism,

---

* Finley Peter Dunne may have liked individual priests, but he was suspicious nevertheless: "I have never known an unworthy priest, but the less you have to do with them the better."

and both the assimilationist younger generation and the militant Polish older generation were dissatisfied with what they thought were his frequently devious compromises.

But for all the opposition and the conflict, Bojnowski held the parish together, built grammar schools, high schools, cemeteries, orphanages. He also published a weekly newspaper in which he lectured the Polish government, the American government, and the Catholic church on how they ought to discharge their responsibilities.

Bojnowski was not only a fighter; he was an elegant and charming man who spoke not only for the Poles to the rest of the community of New Britain, but became one of the community leaders of the city. He won the grudging admiration, respect, and even friendship of both his Yankee Protestant and German and Irish American fellow citizens. He was not a man without flaw, however. He was on occasion guilty of riding roughshod over the rights of his parishioners.*

I do not mention Bojnowski to assert that those who led the neighborhoods were perfect and that no injustice was ever done there. Rather, in the praxis of the neighborhood remarkable skills at coalition building, community development, and reconciliation were necessary, and a generation of men exercised these skills, however narrow or rigid or autocratic they may have been. To write them off as tyrants is cheap and easy; to judge their theology by the theology of today is to make them look old-fashioned and reactionary. Men like "Father B" (as Bojnowski like to be called) are symbols of a critical era in American immigrant history. They were the leaders who gave substantial direction to the course of that history. Failure to understand the immensity of the challenge they faced, the importance of the work they did, and the

---

* A recent article in *Polish American Studies* by Anthony Kuzniewski gives a different view of the Polish National Catholic Church. Kuzniewski summarizes a yet untranslated book by a Polish Marxist historian, Hieronim Kubiak, *Polski Narodowy Kosciol Katolicki w Stanach Zjednoczonych Ameriyki w latach 1897–1965* [The Polish National Catholic Church in the United States of America, 1897–1965] (Krakow: Polish Academy of Science, 1970). There is a substantial irony in that a Polish Marxist concludes that the Polish National Catholic Church has become purely American despite the fact that it was set up originally to protect Polish immigrants from the Americanization pressures of the Irish hierarchy. Kuzniewski, the American non-Marxist Polish reviewer, is in the curious position of saying that neither the Polish National Catholic Church nor the Polish American community have assimilated nearly as much as it appears to Kubiak. "It is important to bear in mind that the same process, viewed from Chicago rather than from Warsaw, could be seen in a different light. For to maintain that declension from the Old World outlooks and manner of life has taken place is not to deny the remarkable tenacity of traditional Polish elements among the Polish Americans. The emphasis depends upon the point of view."[15]

The irony is compounded by the fact that the "first scholarly analysis of the historical and social implications of the Polish National Catholic Church," was written in Polish, published by the Polish Academy of Science in Warsaw, and grew out of research by neither a Roman Catholic nor a Polish National Catholic, but a Marxist historian.

remarkable skills they applied to their task severely curtails our knowledge about what went into the creation of American cities.

A different approach was that of Father Vincent Barzynski, who in 1874 founded the famous Stanislaw Kostka in Chicago, which by 1899 had become the largest parish in the world and was celebrated in Thomas and Znaniecki classic, *The Polish Peasant in Europe and America*.[16] The "Polish downtown" or "little Poland" or "Stanislowowo," as the Barzynski area was called, offered a full range of social services and cultural activities dedicated essentially to Americanizing the immigrant. Polish culture was to survive in some form, but it was thought important by Barzynski and his colleagues in the Resurrection and Holy Cross orders, which serviced the Stanislowowo, that the immigrants became "good yankees."

Nonetheless, one need only tour the Stanislowowo now, long after immigration has come to an end (the neighborhood is now becoming predominantly Puerto Rican), to see the grammar schools, high schools, community centers, and fraternal organizations which denote a tremendously strong neighborhood.

At the opposite extreme to Barzynski was Francis Hodur, a Polish priest who broke with Catholicism and founded the first national Catholic church, which made limited inroads on Roman Catholicism in both Chicago and Pittsburgh. From Hodur's viewpoint, both the Barzynski and Bojnowski approaches ceded far too much to American culture and to American Irish ecclesiastical structure. The Polish National Catholic Church was never large enough or strong enough to develop the kinds of neighborhoods over which Bojnowski and Barzynski presided.

The neighborhood was also the educational base, particularly in the parochial school. In some communities Catholics so dominated the public school system that it was scarcely necessary to build a parochial school (particularly in the Boston area). In Holy Family parish on the west side of Chicago in the years between the Civil War and the great Chicago fire, the legendary Father Damen (after whom Damen Avenue, Chicago, is named) had five schools within his parish boundaries, including one "college," a four-year preparatory and two-year collegiate institution from which eventually Loyola University grew. The immigrants learned quickly that education was the way to make it in American society. To send their children to secondary and then, in the next generation, to higher education became the dream of every parent.

The neighborhood, then, was a receiving station from the Old World and a transmitting station for the New. It was an institution of Americanization from which immigrants could begin to find their way into the political, economic, social, and educational life of the new country, not as isolated individuals but as members of a strong, supportive community, strongly committed (perhaps implicitly) to the pluralistic notion that anyone could become an American, and it was not necessary to give up one's traditional loyalties, heritage, family, or church in the process.*

The New World admitted strangers but was not particularly friendly toward them. The neighborhood was a bastion, a bulwark, an outpost, and from the point of view perhaps of the host society, also something of a Trojan horse.

It is of critical importance to realize that there was almost no such thing as a homogeneous ethnic neighborhood, at least not after the Civil War. One can stand in Bridgeport and see within three blocks five church steeples—the Polish, the Czech, the German, the Lithuanian, and the "Irish." The "Irish" church was a territorial parish which has more recently become known as the "Italian" parish. All Saints, St. John Nepomucene, St. Anthony, and St. Barbara all represented different ethnic enclaves within the same neighborhood. (On one street corner in the city of Rome, New York, there are four different churches with a fifth a half block away.) Even in neighborhoods where the density of national churches was not so great, there were still almost always at least two and usually more ethnic groups cooperating, competing, struggling, getting along. Sometimes there was harmony, sometimes confrontation, sometimes alliance. Even in Bridgeport the Irish were forced to share the neighborhood with Germans.

From the beginning the neighborhood had to absorb diversity, the diversity of the Old World and the New, of the recent immigrants (the "greenhorns") and the old-timers, and finally the diversity of styles, religion, and customs within the ethnic collectivities themselves. The single-group ethnic neighborhood is mostly a myth, though there were certainly areas of the city where one group was more concentrated and dominant than others. If one looks at census material even in the years around the turn of the century, one will find few homogeneous neighborhoods. If the neighborhood was not ethnically diverse from the beginning, it soon became so. The near west side of Chicago, for example, was "Little Italy" and also "Greektown." The Irish had to share the south side of Chicago with Germans and Swedes, and later Italians, Poles, Lithuanians, and blacks. There was not always peace and fraternity among these diverse groups, as Dooley's account of the aldermanic race against Schwartzmeister makes clear. The tragic 1919 race riot is somber evidence that pluralism didn't always work. Perhaps

* It was said of "old" Father Dorney (as opposed to "young" Father Dorney, who baptised me in 1928) that when the hearse carrying his corpse to Holy Sepulchre Cemetery arrived at the graveyard, the last carriage of the procession had not yet left St. Gabriel's.

the amazing thing, though, is that there were so few racial and ethnic riots in the big cities after 1870.

Were the neighborhoods good places to live? Most of those who lived in them thought so, and still think so. In the Taylor Street and Bridgeport neighborhood of Chicago, for example, there are many people who could easily move to more affluent communities but will not do so because of the depth of their affection for the neighborhood. Those of us who may be two or three generations removed from the immigrant experience but who have still lived and worked in neighborhoods are perfectly willing to testify that they can be warm, supportive places. A good neighborhood becomes part of your life; you can never really leave it behind. Even in the 1970s many young people decide to move back to their neighborhoods if they can. A neighborhood is something less than an extended family, but it is something infinitely more than an undifferentiated suburban housing tract.

The neighborhood has received a bad press from those Catholic intellectuals who left it—usually feeling they were driven out. If one reads of James Farrell's *Studs Lonigan*, one need not bother with any attacks by lesser writers on the narrow, wretched parochialism of the neighborhood as a place of "spiritual poverty," as Farrell puts it.

Of course, the intellectual had to question the institutions of the neighborhood. Unless he was an extraordinarily well-balanced and mature person, he had to fight the neighborhood and eventually flee from it if he was to pursue his intellectual vocation. However, it is my impression that by the 1930s, and certainly by the 1940s, the situation had changed considerably. Intellectuals still left, but more, I think, to work out their own psychological problems than because the neighborhood would not tolerate them. By the 1950s neighborhoods became quite proud of the Ph.D.'s, college professors, writers, and artists living on their streets.

The neighborhood, unlike the typical suburb, has the full range of age and occupational categories. The young, middle-aged, and the old; the toddler, school child, teenager, young unmarried, the just-married, parents, and the aged matriarchs and patriarchs of the family all live there. In the neighborhood most people are Democrats, but there are Republicans, too, complete with their own precinct captains and ward committeemen. Most are honest, some are not. Virtually all honor the traditional symbols of religion and family life in theory, but in practice they range from saint to atheist and from the faithful spouse to the one who "plays around." Few in the neighborhood are very rich or very poor, but most neighborhoods are comprised of both blue-collar and white-collar, both college-trained professional and manual workers. (Even a well-to-do Irish neighborhood like Beverly has, I believe, a wider range of income than the newer suburbs to the west or to the east.) Some people work in the neighborhood, some not; some run their own businesses, some work for large companies. Some live their lives for their children, others expect the children to live their lives for their parents. There are different religions and different ethnic groups, all free to practice their own customs and beliefs. While there are very few radicals and perhaps a few more reactionaries, there is still a wide range of political opinion in the neighborhood; no one is under moral constraint to repeat a fashionable party line.

The neighborhood has room for cranks and crabs, for the physically and mentally handicapped, and a diversity of eccentric personalities. Every real neighborhood, as a matter of fact, has to have at least one crank (the man who raises guinea pigs in his garage), one crab (he calls the police every time kids go by his house), and one or more haunted house. If there is no haunted house, there is no neighborhood. Mr. Ed the Weatherman (he remembers the weather for every day of the past forty years), Sam Sam the Candyman (he gives candy to all the kids on the street), and the Cat Lady (she lives in a twisted, unpainted house with a vast number of felines) are more than just neighborhood characters; they are evidence that the neighborhood is strong enough in its patterned relationships, its customs and folkways, to tolerate not only diversity but some eccentricity and a bit of the bizarre.*

If you want privacy in the neighborhood, you can have it; if you want community, you can have that, too. One young man who lived in an upper-middle-class Irish neighborhood told me that for three weeks after his wife was suddenly hospitalized with an almost fatal illness, he came home from work to discover a different neighbor preparing the family supper every night. Gemeinschaft like this is hard to come by. Still, if you want to be a recluse, that's all right, too. The kids may make up stories about you, but if you are determined in your reclusivity, you won't give a damn what they say anyhow.

The neighborhood also generates tremendous loyalty. If you encounter someone from your neighborhood halfway around the world, or receive a letter from someone you knew in the neighborhood thirty years ago, there is a compulsion to respond as though he were a close friend, whether or not the relationship within the neighborhood was particularly friendly. Neighborhood people are special. You feel a certain obligation toward them and they to you. It is easy to see in this mutuality of obligation a continuation in the urban environment of the old peasant

---

* In the neighborhood you are not likely to get hassled to sign petitions for worthy causes. As one young mother put it after she moved out of a university community back into her old neighborhood, "I'll be damned if my children go through life feeling guilty because they don't want to sign a petition every time they go into the Co-op."

loyalties of village and clan. Surely the loyalties a neighborhood generates are less formalized and ritualized than those to an extended family in a farming community in the west of Ireland, let us say. In a way, however, the neighborhood ties can be even more powerful because they are to some extent a matter of free choice.

Of critical importance to the neighborhood are the places where people can gather. The front porch or the door stoop is the primary gathering place. Absence of both of these structures in the suburbs may be one of the reasons why communal life deteriorates so quickly there. I simply cannot believe that something as important to generations of American society as the front porch was eliminated without resistance in almost all construction since 1945. Another important institution which is rare in the suburbs is the alley. It is only secondarily a place for garbage trucks, primarily a communication channel for neighboring. For reasons not altogether clear to me, an alley links people together rather than separating them. In the neighborhood you are close to the people who live across the alley from you, closer even than to those who live across the street. In the suburbs, however, you interact with the people across the street. Perhaps all this has to do with the location of parking lots and garbage cans.

The secondary meeting places in the neighborhood are the candy store, the drugstore corner, the pool hall, and the tavern. Most of these have been eliminated from the suburbs. In Beverly, where I lived for ten years, the drugstore was there, but many of the quasi-suburbanite adults who lived in Beverly felt that adolescents "hanging around" that corner to be an abomination. They did everything they could to destroy the institution. Unfortunately, they were successful. In many Catholic parishes—where the pastor's disposition permits—schoolyards make a good place for "hanging around." In the occasional extraordinary parish, the rectory basement serves that purpose admirably.

The clean, neat, antiseptic suburbs, laid out by city planners or built haphazardly by developers, may have all kinds of park and education facilities, but they rarely provide young people with any place to "hang out." Small wonder they take to hot rods and drugs. The point is that the front porch, the door stoop, the drugstore corner, the alley are all institutions that contributed to the toleration of diversity, if not its reinforcement. Their disappearance from suburban construction was a strong factor toward impelling homogeneity.

There is less age and sex stereotyping in the neighborhood than there is in the one-class, one-age suburb. While the neighborhood has standards about appropriate behavior for people of various ages of both sexes, there is a wide latitude in the application of these standards, precisely because there are present in most neighborhoods different generations and different occupational groupings. For example, a woman may work if she wants to (in the lower-middle-class and working-class neighborhoods women have had jobs—if not careers—for a long time without ever needing a liberation movement to legitimate such occupation). But a woman can also be "just a housewife" if she wants, an option scarcely left anymore to a woman in a university community. Nor is there in the neighborhood anything nearly so rigid as the status structure imposed on the university community by the tenure and promotion system, or by salary, automobiles, fur coats, and foreign vacations in the upper-class suburb. In the neighborhoods I know best, the people with the most money by no means have the highest status, and the highest-status people are not necessarily wealthy.

So the neighborhood is a place of variety—Old World and New, Protestant, Catholic, and in some cities Jewish; Irish, Italian, German, Polish, Lithuanian; devout, reverent; drunk and sober, success and failure, honest and dishonest. The neighborhood can be rigid, but it can also be warm and supportive. It can be parochial yet exciting. It gives a lot but makes a lot of demands. You can never forget it no matter how long you are away from it.

I am a bit uneasy at the conclusion of this chapter. In virtually all the other chapters I think my data are incontrovertible. One may choose not to believe them, but the data still represents the hard facts of the most sophisticated modern empirical research. That kind of hard fact about neighborhoods simply does not exist. (There is, however, some suggestive data on the subject which I shall present in Chapter 13.) The depth of my personal attachment to the neighborhoods I have known and loved, and my horror at the casual way administrators, planners, and academics dismiss them, leads me toward poetry and encomium on the subject of neighborhoods. If one wishes to understand American Catholics, one will have to understand that there are large numbers of us who share this propensity. It may be second-rate and mediocre of us, it may be abysmally particularistic and regressive of us, as some Catholic intellectuals have suggested; it may be that we will have to give up this nonideological localism if we wish to be fully accepted into American society, but we will probably persist in our unreasoned attachment to our neighborhoods. (Paradoxically enough, some New Leftists seem to think that neighborhoods are the wave of the future.)[17] Much more comparative research on neighborhoods simply must be done.

REFERENCES

1. Norman M. Bradburn, Seymour Sudman, and Galen L. Gockel, *Racial Integration in American Neighborhoods: A Comparative Survey* (Chicago: A National Opinion Research Center Report, 1970).

2. Herbert Gans, *The Urban Villagers* (Glencoe, Ill.: The Free Press, 1962).

3. Gerald D. Suttles, *The Social Order of the Slum* (Chicago: University of Chicago Press, 1968).

4. James J. Duran, "The Ecology of Ethnic Groups from a Kenyan Perspective," *Ethnicity* 1 (April 1974): 43–64.

5. Jay P. Dolan, *The Immigrant Church: New York's Irish and German Catholics, 1815–1865* (Baltimore: Johns Hopkins University Press, 1975), p. iii.

6. Ibid., p. 288.

7. Ibid., p. 294.

8. Ibid.

9. Nelson J. Callahan, "Culture as a Factor in Matrimonial Consent" (address to the Midwest convention of the Canon Law Society, Cleveland, Ohio, April 22, 1975).

10. Oscar Handlin, *The Uprooted* (Boston: Little, Brown and Co., 1951).

11. William I. Thomas and Florjan Znaniecki, *The Polish Peasant in Europe and America*, 5 vols. (Chicago: University of Chicago Press, 1918).

12. Georg Mann, "Appreciation," *Eire-Ireland* 9 (Autumn 1974): 124–125.

13. Charles Shanabruch, *The Catholic Church's Role in the Americanization of Chicago's Immigrants, 1833–1928*, Ph.D. diss. (Chicago: University of Chicago Press, 1975).

14. Daniel Buczek, *Immigrant Pastor* (Waterbury, Conn.: Hemingway, 1974).

15. Anthony Kuzniewski, "The Polish National Catholic Church—The View from People's Poland," *Polish American Studies* 31 (Spring 1974): 34.

16. Thomas and Znaniecki, *Polish Peasant in Europe and America*, vol. 5.

17. David Morris and Karl Hess, *Neighborhood Power* (Boston: Beacon Press, 1975).

CHAPTER
12

A Matter of Values

THE WEEK before I began to write the first draft of this book, an article entitled "Catholic Loyalty Versus Protestant Morality" appeared in the *Christian Century*.[1] It was part of a tendentious and fundamentally silly argument between the Catholic commentator Michael Novak and one Erling Skorbin as to whether there are fundamentally different approaches to moral issues and, through moral issues, to political policy questions such as racial tension, busing, and political corruption. The argument is silly because neither one of the writers feels the slightest obligation to be burdened by empirical evidence for his broad and sweeping generalizations about the different moral styles of Protestants and Catholics. Still, the controversy does reflect some widely held assumptions on both sides of the religious fence. Protestants are much more likely to be concerned about issues of personal honesty, political corruption, gambling, alcohol consumption, and, more recently, violence on the TV screen, for instance. Catholics, it seemed, are a good deal more concerned about both loyalty to friends and family on the one hand, and stern sexual morality on the other.

There has been only one empirical attempt to test this widespread "picture," as far as I know. Robert McNamara, Department of Sociology, Loyola University in Chicago, studied college students in a number of eastern colleges and discovered that the Catholic students were much more likely to have stern sexual mores and to be relaxed about questions of personal honesty, while Protestants were much stricter in their notions of respect for the truth and for other people's property, and much more relaxed about sexual issues.[2] The McNamara study, however, is a rather thin base for broad generalizations about the moral styles of the various American denominational groups.

However, in the middle 1960s the National Opinion Research Center (NORC) conducted for one of the major TV networks an extremely innovative study of moral attitudes and behavior.* The survey instrument consisted of twenty "moral vignettes," examples of the moral decision-making problems that arise, with greater or lesser frequency, in the lives of most Americans (see following list). The respondents who were presented with each of the moral vignettes were asked if they would do what the decision maker in the vignette had done. They were then asked whether they thought the decision was wrong, and how wrong it was. Finally, two sets of extenuating circumstances were provided, and the respondent was given an opportunity to judge whether extenuating circumstances might excuse an action which in principle would have been judged immoral.

LIST OF ITEMS
(not in strict questionnaire format)

1. *Dress*

A woman is asked by her daughter if she likes the new dress the daughter is wearing. The mother thinks that the dress does not fit well and the color is not flattering, but she tells her daughter the dress looks fine anyway.

If you were the mother, would you have said that?

Extenuating circumstance 1.

Suppose the mother knows that her daughter is sensitive to criticism and will sometimes cry or argue if her taste is challenged. She tells her daughter the dress is fine. Under those circumstances, is it right or wrong for the mother to say what she did?

Extenuating circumstance 2.

Suppose the mother knows that the young girl has made the dress herself and is very proud of what she has done. It is the first dress she ever made and she worked on it for weeks. She tells her daughter the dress is fine. Under those circumstances, was the mother right or wrong to tell the daughter what she did?

2. *Affair*

Mr. Martin has been married for 10 years and has two children. He met Miss Brown, an attractive, unmarried young woman with whom he drifted into an intimate relationship. There was no possibility of anyone finding out about it since his job gave him a good excuse to be away from home. The young woman seemed happy with the situation and had no desire to have Mr. Martin get a divorce and marry her.

As you see it, was Mr. Martin right or wrong in getting involved with Miss Brown?

How wrong?

* The results of this study were not published at the time. The data were rediscovered recently by Kevin Ryan, professor of education, Ohio State University. I am grateful to him for making the data available to me for this chapter.

Extenuating circumstance 1.
Suppose for the last several years his wife had been cold to him and had argued with him continuously. In that case, was Mr. Martin right or wrong?
Extenuating circumstance 2.
Suppose since meeting Miss Brown he had actually become a better husband and father since he was no longer depressed and he and his wife didn't argue so much. In that case, was Mr. Martin right or wrong?

3. Bribe

A salesman who uses his car to make all his calls is driving down a deserted street in a strange town early one morning. He speeds at 55 mph in a 35 mph zone. He is stopped by a policeman who tells him that the fine is $40. He offers the policeman $10 to let him go.
If you were in that situation, would you have offered the policeman $10 to let you go?
How seriously wrong is it?
Extenuating circumstance 1.
Suppose the policeman suggests by his behavior that he expects to be bribed. In that case, was the man right or wrong?
Extenuating circumstance 2.
Suppose the policeman tells the man he will have to wait until afternoon to be tried by the justice of the peace. The man knows that he can't take the time to wait all day, because he has an early appointment with a customer who is buying thousands of dollars worth of his company's product. In that case, was the man right or wrong to offer the policeman $10 to let him go?

4. Money

A young office clerk spots a $50 bill on her boss's desk. She knows she can take the money without anyone knowing. She puts the money in her purse.
Would you have done that?
Was the girl right or wrong to take the money?
Extenuating circumstance 1.
Suppose the boss is a rich, old grouch who has made many unreasonable demands on the girl, and pays her only a small salary. In that case, was she right or wrong to take the $50 bill?
Extenuating circumstance 2.
The girl gives the $50 to her widowed mother, who needs it. In that case, was she right or wrong to take the $50 bill?

5. Grades

At a small high school the principal worked hard to get each student into the college he wished to attend. One boy had done poorly in his final year, and the principal believed that his low grades would cause him to be rejected by the college the boy had chosen. He decided to raise the boy's grades so that he would be accepted by the college.
If you were the principal, would you have done this?
Was the principal right or wrong?
Extenuating circumstance 1.
Suppose the boy had received good grades on his college entrance exams and the principal was sure the boy would do well in college, but the principal knew that this particular college placed the most importance on high

school grades. In that case, do you think it was right or wrong for the principal to raise the boy's grades?

Extenuating circumstance 2.

Suppose the principal also knew that the boy had been sick in the hospital for several months and this was why his grades were lower in the senior year. In that case, would you say what the principal did was right or wrong?

6. Test questions

Sally and Jane, who are seniors in high school, are good friends. Sally is taking an exam in math in the afternoon, while Jane has already had the same exam in the morning. Sally asks Jane to tell her what the questions on the exam are.

If you were Sally, would you have asked for the questions?

Was Sally right or wrong in asking what the questions were?

Extenuating circumstance 1.

Suppose that other students in the class will be getting the answers from someone else, and Sally feels it is really the instructor's fault for not changing the test questions for the afternoon class. In that case, is Sally right or wrong in asking what the questions were?

Extenuating circumstance 2.

Suppose Sally is a good student and has worked hard in math all year, but is still having great difficulty with it. Her final mark depends on the test she is taking this afternoon. If she doesn't pass the course, she will not graduate. In that case, is Sally right or wrong?

7. Bricks

A construction worker takes home from the place where he is working enough bricks and wood to build a porch for his own home. He takes a small amount each day, and the inventory system is so loose that he is sure the materials will not be missed.

If you were the construction worker, would you have done that?

Was the construction worker right or wrong to take the bricks and wood? How wrong?

Extenuating circumstance 1.

Suppose he knows that the purchasing agent has bought more bricks and wood than are really needed on the job just to be sure that the work is not stopped due to lack of material, and whatever is left over after the building is finished will be dumped or buried. In that case, was the worker right or wrong to take the bricks and wood?

Extenuating circumstance 2.

Suppose the man plans to use the bricks and wood to build a shelter at his son's boy scout camp. In that case, was the worker right or wrong to take the bricks and wood?

8. Fender

A woman hits the front fender of a parked car and puts a dent in it. She's insured but rather than take the time to stop and report the matter, she just keeps going.

If you had hit the parked car, would you have done that?

Was the woman right or wrong to do what she did? How wrong?

Extenuating circumstance 1.

Suppose the car she hit was double parked and is a five-year-old model which already has several scratches and dents. In that case, is the woman right or wrong to keep going?

Extenuating circumstance 2.

Suppose the woman knows her insurance company will cancel her policy if the accident is reported, and she will have to pay a high rate for new insurance. In that case, is the woman right or wrong to do what she did?

9. *Charges*

The man whose fender was dented has $50 deductible insurance—that is, his insurance company pays for everything over the first $50 of the bill. He takes it to a garage to be fixed, and the garage owner says the job will cost $75 but that he will put $125 on the bill, so that it doesn't cost the man anything. The man agrees to the garage owner's suggestion.

If you were the man, would you have agreed to the garage owner's suggestion?
Was the man right or wrong? How wrong?

Charges 2.

While the garage owner is examining the car, he notices a little dent in the door. The man says it was done a couple of months ago by his son's bike. The garageman says he can fix it at the same time and just put it on the bill to the insurance company, too. The man tells him to do that. Is the man right or wrong?

Charges 3.

A couple of years ago the garage owner would not have padded the bill for his customers, but he felt he was losing business to garages which did so, so he too started doing it. Is the garageman right or wrong?

10. *Clip*

While walking through a busy railroad terminal on the way to meet a friend coming in on the train, a man sees an unusual money clip on the floor with $5 in it. He picks it up and instead of turning it in to the Lost and Found, keeps the money.

If you happened to spot a money clip with $5 in it, would you pick it up and keep it?
Do you think the man was right or wrong to pick up and keep the $5? How wrong?

Extenuating circumstance 1.

Suppose the person who dropped the money was very well dressed and carrying an expensive suitcase. It is clear that $5 couldn't mean very much to such a person. In that case, was the man right or wrong to keep the money?

Extenuating circumstance 2.

Suppose the man means to take the money clip to the Lost and Found office, but in the excitement of meeting his friend he forgets about it until he gets home. He lives miles from the station. In that case, is he right or wrong to keep the money?

11. *Tax*

A man claims his mother as an income tax deduction even though he only contributes 40 percent to her support, instead of the 50 percent required by law. The rest of her income comes from Social Security payments, so no one can really claim her as a deduction.

If you were in that situation, would you claim the mother?
Was the man right or wrong to claim his mother as a deduction? How wrong?

Extenuating circumstance 1.
Suppose the government is not paying his mother certain Social Security benefits that he is convinced she is entitled to. In that case, is he right or wrong?

Extenuating circumstance 2.
Suppose the man is having financial problems, has a large family, and finds it hard to give his mother even the small amount he now gives her. In that case, is he right or wrong?

12. *Sweater*
A woman receives the coat which she has ordered from a large mail order house. In addition to the coat there is also a sweater which she did not order and for which she is not charged. She keeps the sweater.

If you were the person to whom this happened, would you keep the sweater?

Was the woman right or wrong to keep the sweater? How wrong?

Extenuating circumstance 1.
Suppose she knows that it would cost a big mail order house more to correct the error than the sweater was worth. In that case, was the woman right or wrong to keep the sweater?

Extenuating circumstance 2.
Suppose she then takes the sweater and contributes it to the local church rummage sale. In that case, was she right or wrong to keep the sweater?

13. *Quiz show*
The producer of a television quiz show is visited by the sponsor and told that the program must get higher ratings—that is, more people must watch it. The sponsor and producer know that the only way to do this is to give the better contestants the answers and keep them winning and thus build interest in the program. The producer rigs the show.

If you were the producer, would you have done that?

Is the producer right or wrong to rig the show? How wrong?

Extenuating circumstance 1.
Suppose the producer believes that people watch television for entertainment, and that they get more enjoyment from seeing interesting contestants win and he sees no way in which the audience is harmed. In that case, is the producer right or wrong?

Extenuating circumstance 2.
Suppose that the sponsor and the network tell the producer that if he refuses to cooperate, the show will go off the air, and he will lose his job. In that case, is the producer right or wrong?

14. *Dinner*
A man and his wife are traveling along a turnpike and turn off at a small diner to get supper. When the man pays his bill, the owner gives him five dollars in change instead of one. The man puts the money in his pocket and drives away.

If that happened to you, would you have done what the man did?

Was the man right or wrong to keep the money? How wrong?

Extenuating circumstance 1.
The couple had received very poor service in the diner, and the food which they ordered was cold and tasteless. In that case, is the man right or wrong?

Extenuating circumstance 2.
The man later gave the five dollars to the American Cancer Society. In that case, is the man right or wrong?

15. *Towel*
A family spends the night at a motel while they are on vacation. They notice that the maid has left a couple of extra towels with a nice design on them. The next morning as they leave, they put one of the towels in their suitcase.
If you were in a motel, would you have taken a towel?
Was the family right or wrong to take the towel? How wrong?
Extenuating circumstance 1.
Suppose the family has received poor service at the motel and had to pay more for their room than they expected to, and they felt cheated. In that case, was the family right or wrong to take the towel?
Extenuating circumstance 2.
Suppose the father says that the motel expects guests to take towels and that this is good advertising for them since their name is on the towels. In that case, was the family right or wrong to take the towel?

16. *Book*
A college student sees a reference book in the library which must be used in the library only and which she will often have to refer to for a special report. While the librarian's back is turned, she puts the book in her briefcase and walks out. She plans to return the book when her report is finished.
Would you have done that?
Was the student right or wrong to take the book? How wrong?
Extenuating circumstance 1.
Suppose she can tell from the dates in the book that no one had used it in several years. In that case, was the student right or wrong to take the book?
Extenuating circumstance 2.
Suppose she is writing an important special report and needs the book very badly. She has no other way of getting it. In that case, is she right or wrong to take the book?

17. *Votes*
During a political campaign a candidate for Congress pledges his support for a new dam over a local creek. He does not believe the dam is really needed, and he has no intention of working for it if elected, but he knows this will get him a lot of votes.
If you were the candidate, would you do this?
As you see it, was the candidate right or wrong? How wrong?
Extenuating circumstance 1.
Suppose that several of his largest financial backers favor the dam, and he can't win without their support. In that case, is the candidate right or wrong?
Extenuating circumstance 2.
Suppose the candidate believes he can do a great deal of good if elected, and he is convinced that his opponent is a crooked politician. In that case, is he right or wrong?

18. *Half-fare*
A man takes his 12-year-old son to the bus depot to buy him a ticket to travel across country to visit his grandparents. The boy is small and can easily pass

for less than 12. Children under 12 pay half-fare. If he rides for half-fare, he saves $40. The father buys the half-fare ticket.

> If you were the father, would you buy the half-fare ticket?
> Was the father right or wrong to buy the half-fare ticket? How wrong?
> *Extenuating circumstance 1.*
> Suppose the boy turned 12 only a few days ago, and also the bus company had just raised its fares. In that case, was the father right or wrong?
> *Extenuating circumstance 2.*
> Suppose the family could not afford to send the boy to visit his grandparents at all if the ticket cost $40 more. In that case, was the father right or wrong?

19. Secretary
A salesman calls to visit the president of a company. The president's secretary knows her boss is busy working on a report and tells the salesman that he is out of town.

> If you were the secretary, would you have told the salesman that?
> Was the secretary right or wrong to tell the salesman that her boss is out of town? How wrong?
> *Extenuating circumstance 1.*
> Suppose the salesman has been very annoying and persistent and has frequently disturbed her boss. In that case, is the secretary right or wrong?
> *Extenuating circumstance 2.*
> Suppose also that the president has given the secretary instructions that he does not wish to be disturbed and to tell all salesmen that he is out of town. In that case, is the secretary right or wrong?

20. Embezzle
A bookkeeper working for the local city government saw a way in which he could alter the books and take $40 without its being noticed. He takes the money.

> If you were the bookkeeper, would you alter the books to take $40?
> Was the bookkeeper right or wrong to do this? How wrong?
> *Extenuating circumstance 1.*
> Suppose he had long been underpaid and had recently been denied a small raise he felt he was entitled to. In that case, was he right or wrong to alter the books?
> *Extenuating circumstance 2.*
> Suppose his wife was seriously ill and he spent the money to buy her medicine. In that case, was he right or wrong to alter the books?

One need only look at the preceding vignettes to see how difficult and complicated are the questions of moral decision making once one removes it from the abstract and puts it into concrete practical cases. The line between corruption and flexibility is thin and, equally, the line between principle and rigidity can be thin. Is it a matter of high principles, or is it rigid and unfeeling inflexibility, for example, that would lead a mother to tell a sensitive daughter that a dress her daughter is proud of does not look "fine" on her? Yet we note that better than four-fifths of the American population claim that it is wrong to give a civil and conventional answer to the question (Table 12.1).

TABLE 12.1

Moral Decisions by Religious Denomination

VIGNETTE	% WOULD NOT DO/SAY			% WRONG			% VERY WRONG			EXTEN. 1 WRONG			EXTEN. 2 WRONG		
	PROT.	CATH.	JEW	PROT.	CATH.	JEW	PROT.	CATH.	JEW	PROT.	CATH.	JEW	PROT.	CATH.	JEW
Dress	84	86	78	84	79	78	23	14	14	80	75	64	49	41	32
Affair	–	–	–	98	97	100	86	89	83	90	86	83	86	82	80
Bribe	88	87	72	96	96	92	67	66	68	91	86	91	87	81	73
Money	98	98	97	97	99	100	93	92	94	97	96	100	96	93	100
Grade	87	85	86	93	93	99	67	66	53	84	80	69	68	64	63
Test questions	69	62	59	89	86	89	44	45	53	92	90	93	89	89	88
Bricks	92	86	86*	98	97	94	72	60	71	88	78	86	93	87	91*
Fender	95	94	87	99	99	100	70	65	58	98	95	97	97	97	94
Charges	73	65	41*	93	92	89	62	49	31*	91	90	86	92	90	77*
Clip	69	59	47*	80	72	66*	64	58	53*	99	95	100	82	66	75*
Tax	79	70	53*	87	85	72*	66	52	35*	90	81	92*	90	84	87*
Sweater	87	88	80	93	92	89	47	42	25*	91	90	97	94	89	87*
Quiz show	84	82	86	94	93	97	72	69	66	89	88	91	89	84	80*
Dinner	89	90	92	97	96	100	58	56	44	96	95	100	94	91	94
Towel	87	77	78*	97	95	97	46	31	23*	98	96	100	92	87	88
Books	98	97	100	98	96	100	51	46	55	97	97	100	94	90	94
Votes	87	81	72*	97	95	94	74	72	53	96	89	91*	85	76	68*
Half-fare	65	47	36*	90	81	86*	44	34	19*	92	85	74*	89	82	71*
Secretary	68	54	59*	81	71	78	28	25	14	74	66	57*	41	25	18*
Embezzle	60	58	40*	93	94	84	87	87	71	–	–	–	–	–	–

* Responses among denominations significantly different.

Similarly, is it wrong to take money to keep one's wife alive (as in the final case)? Some three-fifths of the American population think that it is. And yet I wonder if any ethical philosopher or theologian—Protestant, Catholic, Jew, or humanist—would not take a much more benign view of the decision making on that question. Thus even before we can take a detailed look at the data, we must recognize that the question is not one of moral inferiority or superiority; it is difficult to rank population groups in their response to the moral vignettes in order to separate out moral flexibility and sophistication from moral strength and weakness. It is clear that the findings must necessarily be ambiguous. The three American denominational groups* may handle moral cases differently, and that fact is worth knowing, but it does not follow that a stern propensity to give a morally "correct" answer is necessarily evidence of moral excellence. It could just as easily be a sign of moral rigidity.

The first observation we might make about the summary data is that the three major religious groups are in agreement for the most part in their responses to the moral vignettes. In the majority of the cases there are no statistically significant differences among the three groups as to whether the respondent himself would make the decision described in the vignette. In nine of the cases religion does predict a significant difference in the respondent's own decision making; but only in four cases is there a statistically significant difference among the three denominations in judging whether the action was morally wrong. In only seven cases is there a difference in judging whether the action was "very wrong."

In the first extenuating circumstances column of Table 12.1, religion accounts for statistically significant differences seven times. It is only when one gets to the second extenuating circumstances column that statistically significant differences emerge in the majority of circumstances, a finding not without interest, because the designers of the 1965 survey attempted in this second extenuating circumstance to operationalize Kohlberg's concept of responsible and autonomous morality.[3]

When religion is a statistically significant predictor of differences in moral decision making, the pattern is almost always one of Protestants taking the strictest moral position and Jews the most lenient position, with Catholics generally in the middle. Thus if one argues that Protestants are more moral than Catholics, one must logically conclude that Catholics are more moral than Jews. Or if one argues that Jews are more flexible morally than Catholics, then one must argue that Catholics are more flexible than Protestants. Either case offers little support for the assumption of an "inferior" Catholic morality; nor is there evidence for the argument that Catholics, as members of an "authoritarian" church,

* There was no ethnic question in this study.

are more rigid morally and more unthinking than Protestants. If we can view the subject of the patterns of moral decision making among American denominations with some objectivity, we would observe that when religion does affect moral decision making, Catholics emerge in most cases somewhere in the middle of a continuum, with Protestants and Jews at each end.

Catholics and Jews were somewhat more likely than were Protestants to say they would take bricks from the construction site. Jews were more likely than Catholics and Catholics more likely than Protestants to "cheat" on auto repair charges, to keep a lost money clip with $5 in it, to claim a false income tax deduction, to make false political promises to obtain votes, to try to get a child onto public transportation for half-fare, and to embezzle money to buy a kidney machine for a sick spouse. There was virtually no difference between Catholics and Jews on liberating a towel from a hotel room, with Protestants much less likely to say they would make off with the towel. Finally, in only one instance—the secretary fibbing about her boss' absence from the office—were Catholics less likely than Protestants and Jews to render a "strict" moral judgment.*

In two of the cases in which there was difference among the religious groups as to whether an action was wrong—picking up the $5 money clip and claiming a false income tax exemption—the same Protestant-Catholic-Jew ordering persisted, but Catholics were less likely than either of the other two to think that the secretary's decision about hiding her boss' presence or the attempt to get on public transportation for half-fare was wrong. The traditional Catholic moral teaching saw nothing wrong with avoiding full-fare payment so long as one did not lie in the process.

In *all* the cases where there were differences among the three religious groups in judging whether the moral decision made in the vignette was "very wrong," Protestants were the most likely to judge an act "very wrong," and Jews the least likely. Catholics occupy the middle position with the exception of the "white lie" about the daughter's dress, in which there was no difference between Catholics and Jews. More than 60 percent of the Protestants in the NORC study thought that failing to report the lost $5 bill, cheating on the repair bill, and claiming a false tax exemption were very wrong, while more than two-fifths of them thought that keeping the sweater, lifting the towel from the hotel, and trying to get the child on public transportation for

---

* This is one case for which virtually every Catholic catechism provides an explicit response—legitimating the moral decision of the secretary. Normally, the Catholic catechism would interpret the secretary's response as saying that the boss is not in *to you*, with the last words expressed implicity. Such a response, it was argued, was not an attempt to deceive but merely an urbane form of the brush-off.

half-fare were "very seriously" wrong. Of these cases, the Jewish proportion making the same judgment about the morality of the action was approximately half as high.

Catholics seem to be much more rigorous in their moral judgments than are their own moral theologians, at least in the traditional Catholic theology. Only extramarital affairs would certainly be considered very morally wrong, and the quiz-show rigging, the false tax return, the stealing of construction supplies, and the false appeal to voters would be matters for debate among Catholic moral theologians.

In four of the situations where there is the first extenuating circumstance—the bribe to the policeman, stealing construction material, claiming a false tax exemption, and the making of false campaign promises—Catholics are the most likely to accept the extenuation, while in the other three cases—the daughter's dress, the fibbing secretary, and the half-fare transit rider—the Protestant-Catholic-Jew ordering continues. However, the overwhelming majority of all Americans reject the first extenuating circumstance as an excuse (save for the modest 57 percent of the Jews, who still think the fibbing secretary is acting immorally when she dismisses the annoying and persistent salesman).

The most interesting set of data are those that reveal the substantial differences in moral decision making when the second extenuating circumstance is introduced. Here the differences among American religious groups tend to be very large. In seven of the ten cases, Jews display a more "flexible" approach than Catholics, while in the other three—the damaged fender, the tax exemption, and the mistakenly mailed sweater—Catholics are more "flexible" than Jews. Thus, when a moderately powerful excusing factor is introduced, Catholics and Jews are rather likely to dissent, sometimes quite sharply, from the stern moral judgment of their Protestant counterparts. However, in only two of the nineteen vignettes to which the second extenuating factor is introduced—the daughter's dress and the fibbing secretary—do less than a majority of Americans judge that the action is not morally wrong. Whatever their personal moral choices may be—and our respondents implicitly admit considerable differences between what they would do and what they think is the moral thing to do—there is a surprisingly strong consensus among Americans of all three denominations about what is immoral and what is moral. This is true even when there is a fairly powerful extenuating circumstance which might excuse an otherwise morally unacceptable action.

One ten-year-old survey does not solve the question of whether there are different moral styles among different religious groupings in the United States.* Perhaps all the 1965 study does is to make clear the

* It is a pity that no worthy patron has been found to finance a replication of the 1965 moral value study.

difficulty of measuring such a subtle and complex issue as moral decision making. However, on the critical question of Catholic "corruption" there is no evidence of Catholics being different from Protestants in whether they would offer a bribe to a policeman or approve of "dishonest" political campaigning. In both cases Jews are even less likely than Catholics to "measure up" to the stern moral choices of American Protestants. There is, then, broad consensus among Americans about what is morally right and what is morally wrong, and about the wrongness of certain kinds of immoral behavior. There is some disagreement among members of the three denominations about whether they would engage in such behavior, with Catholics and Jews being more likely than Protestants to acknowledge that they would. In most instances Jews are even more likely than Catholics to admit that they would do something that was "wrong." And this, in all likelihood, is because they are less likely to see it as "very wrong." Since Jews and Catholics are less likely to think that padding a repair bill for an insurance company, taking a $5 money clip, appropriating a sweater sent by mistake, claiming a false income tax exemption, "borrowing" a towel from a hotel, and paying a half-fare for a child of full-fare age are actions which are seriously wrong, they are somewhat more likely to admit that they would do them —perhaps on the grounds that while such actions may be sinful, they are only "small" sins, or, as Catholic moral theology would say, only "venial." Whether the increased propensity to admit that one would "cheat a little" on "small" matters indicates an inferior moral sense among Catholics and Jews, or merely greater realism in self-description, is a question that cannot be resolved with the present data. The similarities of moral style among America's three main denominational groups are much greater than the dissimilarities. To the extent that Catholics are different from the other two denominations, they stand somewhere between Protestants and Jews in their approach to moral problems.

In the complex area of sexual morality, the respondents of the 1965 study differed significantly in their judgments about premarital sex, but the differences were between Jews and Gentiles, there being very little difference between Catholics and Protestants. There was no significant disagreement about extramarital sexuality. About three-quarters of the Protestants and Catholics rejected sex for unmarried couples (Table 12.2, part A). More than half rejected it for engaged couples, while somewhat more than one-third of the Jews rejected such premarital sex for both categories. There is very little difference, however, in the judgments made about premarital and extramarital sex for men and for women—a finding which confirms the large body of research literature which has been unable to unearth any evidence of a double standard in contemporary American life.

TABLE 12.2

Sexual Values by Denominational Groups

A.	MEN			WOMEN		
	PROT.	CATH.	JEW	PROT.	CATH.	JEW
Reject sex for unmarried couples	74	73	31*	78	78	47*
Reject sex for engaged couple	58	58	39*	59	58	38*
Reject extramarital sex	87	89	75	89	90	78

B. PREMARITAL SEX ALWAYS WRONG (PERCENT)

Protestants	
British	29
German	33
Scandinavian	34
Irish	42
"American"	40
Catholics	
Irish	31
German	36
Polish	38
Slavic	33
Italian	30
Jews	11

*Responses among denominations significantly different.

A different wording of the question in more recent NORC General Social Surveys (Table 12.2, part B) shows what may be a dramatic change among Protestants and Catholics in their attitudes toward premarital sex. Interestingly, the Catholic ethnic groups are more flexible on this issue than are the Irish or "American" Protestants. Jews (11 percent) are the least likely to think that premarital sex is always wrong, and British Protestants (29 percent) are next in line; they are followed by Italian Catholics (30 percent) and Irish Catholics (31 percent). However rigid Catholic sexual morality may or may not have been once, it does not seem any more rigid at the present time than the sexual morality of American Protestants.

But perhaps the most striking finding in NORC's recent monitoring of moral attitudes is the basic similarity between Catholics and Protestants on the abortion issue (Table 12.3). In three abortion situations— the danger of the birth of a defective child, threat to a mother's health, and rape—the overwhelming majority of Protestants and Catholics are in favor of legalized abortion. Thus three-quarters of the Irish Catholics approve of legal abortion if there is danger of a defective child or if there is a threat to the mother's health, and 68 percent approve of it

TABLE 12.3

Attitudes on Legal Abortion for White Religio-Ethnic Groups—Metropolitan North Only
(percent accepting)

RELIGIO-ETHNIC GROUP	DEFECT TO CHILD LIKELY	IN CASES OF RAPE	THREAT TO MOTHER'S HEALTH	FAMILY TOO POOR	DOES NOT WANT TO MARRY	NO MORE CHILDREN WANTED
Protestants						
British(218)	93	94	96	65	59	56
German(247)	85	86	92	56	54	46
Scandinavian(74)	78	85	85	57	54	50
Irish(102)	85	89	92	58	57	44
"American"(316)	84	83	91	55	52	48
Catholics						
Irish(113)	76	68	73	36	42	30
German(86)	76	77	82	33	36	38
Polish(71)	74	82	82	44	44	37
Slavic(34)	74	84	88	50	47	37
Italian(146)	99	73	88	54	48	41
Jews(119)	95	98	99	85	86	80

after rape. Despite the considerable pressure brought by Right to Life groups, they apparently do not enjoy anything near majority support even in what is supposedly the most devout Catholic ethnic group, the Irish.

On the other hand, there is only very slight majority support among Protestants for legalized abortion in the other three sets of circumstances—desire to avoid having more children, poverty, and the decision of a pregnant woman not to marry. In two of those cases the "American" Protestant does not give majority support; in one case each, the Irish Protestant and the German Protestant do not give majority support for abortion. Both American Protestants and Catholics, in other words, are not willing to support "abortion on demand," for a majority of both groups reject legal abortion merely as a form of birth control, a means of avoiding more children. And the differences between Protestants and Catholics on the poverty and unmarried mother abortion cases are not especially large.

The controversy over legal abortion, then, is apparently being carried on by minorities on both sides. A broad consensus of Protestants and Catholics favor its legality under certain circumstances, and a majority of both groups do not favor it merely as an alternative form of birth control. The militants on both sides of the controversy are displeased by this finding. The proabortionists will cite the NORC data selectively, omitting the fact that the majority of Catholics and Protestants reject abortion as a form of birth control, whereas the antiabortion group among Catholics periodically inundates NORC with violent letters denouncing the survey findings in the strongest possible language.

To say that the majority of American Catholics favor legalized abortion under certain circumstances does not mean that Catholics are persuaded that abortion is "moral." In the 1974 NORC Parochial School Survey only 6 percent of the American Catholic women (and 7 percent of their husbands) would want abortion for themselves. It would appear that American Catholics are capable of making a distinction between what they think ought to be legally legitimate in a society and what they personally judge to be morally right. The Catholic respondents seem to be saying, in effect, "I would not get an abortion myself, but if others do not think it is immoral, I do not propose to try to force my moral position on them through the laws of the land." It would therefore be a mistake for non-Catholics to conclude that there is a major change taking place among Catholics in their personal respect for the unborn infant. It would be an equal mistake for the Catholic leadership to conclude that there is widespread Catholic support for attempts to obtain blanket legal prohibition for abortion. Both viewpoints, it seems to me, presume that American Catholics are much less sophisticated in their political judgments than the data would suggest.

In summary, there seems to be little difference between American Catholics and American Protestants in their moral judgments, although in some cases Catholics tend to be more "flexible," or "less strict," than Protestants (but also less flexible and more strict than Jews). Differences of opinion between Catholics and Protestants on abortion are not as great as either side on the abortion controversy would make it appear, and the majority of Catholics support the legality of abortion under certain circumstances, though they themselves seem to reject it as a personal moral decision. Finally, if anything, Catholics are less likely than Protestants to judge premarital sex as always wrong, although we do not know from any available data whether there might be a difference between their judgments about morality and their personal moral decisions. As was noted in Chapter 7, there seems to be a rapid change in Catholic moral judgments about sexuality, as well as in Catholic moral attitudes toward birth control. Furthermore, the failure of the church to respond to this change has precipitated a major crisis in organized Catholicism. It does not follow, however, that Catholic sexual behavior will become indistinguishable from that of other Americans, especially since Catholic family values and family structures, as has been noted previously, still seem rather different from those of other

American groups. On the basis of the available data, the most likely prediction for the future would be as follows:

1. Catholics will differ very slightly from anyone else in the society in their practice of birth control.
2. The Catholic divorce rate will increase but will still remain lower than that of Protestants.
3. Catholic family size will diminish to the point where it will be only slightly larger than white Protestant family size.
4. Catholics will grow more tolerant of legalized abortions and may even seek them more often than they did formerly; but they will still overwhelmingly reject abortion as a personal moral choice.
5. Premarital sex among Catholics will increase somewhat, although it will probably be less frequent than it is among some other American groups (especially less frequent than among Jews).
6. Catholics will continue to reject extramarital sex in principle, and probably by a higher plurality than other Americans, though it may diminish somewhat. They will also be less likely to engage in extramarital sex.

The long-run development of Catholic sexual morality in the United States will depend to a considerable extent on how Catholic teaching authority responds to the present crisis of credibility. There is, I suspect, a resource for the church in the strong familial commitment of the Catholic ethnic group, which the teaching authority can mobilize. But to do so it will have to demonstrate much greater flexibility and sensitivity than it has in recent years.

The short-run future for the Catholic sexual ethic would seem to be a mixture of increasing similarity to other Americans in some respects but persistent dissimilarity in others—a projection in keeping with the "ethnicity" model on which this book is based. Differences in Catholic sexual morality will neither be melted down to no differences at all, nor will they be preserved as they were in the past.

There are, then, some differences in moral values between Catholics and non-Catholics. They are not nearly as great as one may have expected. Conflicts may focus on specific moral issues in American society, but they seem to take their origin in deeper matters, such as political style and family structure, as we have noted previously. It may also be, however, that while Catholics and non-Catholics handle specific moral issues in relatively similar ways, they have different attitudes toward their jobs and toward the ultimate meaning of human life (Table 12.4).

There are a number of different scales which purport to measure occupational values. The one I have chosen to use in this chapter is based on a factor analysis done on four years of the NORC General Social Survey.* It has been argued repeatedly that Catholics, particularly eastern and southern European Catholics, lack ambition because of their

* I chose this scale not because of any special intrinsic merit but because it gave more cases for each of the religio-ethnic groups.

TABLE 12.4

Occupational Values Among
*American Religio-Ethnic Groups**
(z scores)[†]

Blacks$_{(345)}$	78
Slavic Catholics$_{(50)}$	32
Italian Catholics$_{(116)}$	09
Polish Catholics$_{(591)}$	05
"American" Protestants$_{(530)}$	01
German Catholics$_{(104)}$	−14
German Protestants$_{(390)}$	−15
Irish Catholics$_{(101)}$	−19
Irish Protestants$_{(144)}$	−20
Scandinavian Protestants$_{(113)}$	−21
British Protestants$_{(368)}$	−33
Jews$_{(83)}$	−43

*Groups ranked according to score on factor measuring importance of income as an occupational goal, as opposed to the inherent meaningfulness of the work.

[†]Deviations from standardized mean.

fatalistic world view and expect their reward in another life instead of this one. The data make it clear that whatever their fatalism and otherworldly view may be, the eastern and southern European Catholics have not been hindered by such values in their struggle for economic success. Their position on the NORC Occupational Value Scale is exactly what one would expect from groups seeking upward mobility in American society. The Slavs, Italians, and Poles are all above the mean on a scale that measures the propensity of a respondent to choose income as an important occupational goal, in preference to the inherent meaningfulness of the work. (Blacks, it is worth noting, are even higher on this scale than eastern and southern European Catholics.) To the extent that the scale can be said to measure an achievement orientation, Catholic ethnics and blacks certainly place economic achievement above other goals in their description of their occupational values. Interestingly, the Catholic groups that have been in the United States for a longer time, the Irish and the Germans, are more likely to stress the inherent meaningfulness of work, and there is little difference between Catholic and Protestant Irish and Catholic and Protestant Germans on this scale.

Having established their credentials as economic achievers, eastern and southern European Catholics are now open to the charge of seeking nothing in life but income. (A reasonable explanation might be that one can only afford to seek inherent meaningfulness in one's work, as do the Jews and British Protestants especially, when one achieves some stable economic success in the society.) It's a heads you win, tails I lose argument. If the Catholic ethnics had scored low on this scale, they would have been accused of lacking ambition; scoring high, they can be accused of lacking refinement and cultivation, despite the fact that some Catholics—the Irish and the Germans—are quite similar to their Protestant counterparts.

There are also differences between Catholics and non-Catholics in ultimate world views. My colleague William McCready has developed a series of vignettes designed to measure how respondents react to critical life situations such as the birth of a handicapped child, a lingering death for a parent, or the discovery that one has a fatal illness. McCready has constructed a typology of ultimate value groupings that is designed to measure the response patterns to these vignettes: "religious optimists" are those who see God providing an easy answer to human crisis situations; "secular optimists" expect the crisis situations to work themselves out, but do not refer explicitly to God as a source of the happy conclusion; the "hopeful" do not deny the horror or the evil in violence and death but see benign powers in the world as ultimately somewhat stronger than the malign ones;* the "pessimists" assume an angry or fatalistic world view; and the "diffuse" are those respondents in whose answers there is no consistent strain or pattern to be found.[4]

Religious optimism and hopefulness are the most frequent Protestant responses, while Catholic responses are most frequently found in the pessimist and diffuse categories (Table 12.5). Hence there is some evidence for Lenski's argument that Catholics tend to be more fatalistic. Clearly they reject both the religious and secular optimistic world views which have particular appeal to Protestants. However, Catholics are far less pessimistic than Jews in the ultimate world view.

Within the Catholic collectivity, the Irish are the most likely to be hopeful—close to the percentage of British Protestants who hold the hopeful view (Table 12.6). The Italians are the most likely to be pessimistic, with some 40 percent of them falling into that category (about the same proportion as Jews). The Irish tend to share the same optimistic and hopeful world view as British Protestants, while the Italian Catholics are much closer to Jews in their pessimism. The differences within the Catholic community in ultimate world view are at least as great as the differences between Catholics and non-Catholics.

* A position McCready and I take to be descriptive of the Catholic Christian tradition.

TABLE 12.5

Proportion Overrepresented or Underrepresented in
Ultimate Value Types by Religious Preference

ULTIMATE VALUE TYPES	PROTESTANT	CATHOLIC	JEWISH	NONE
Religious optimists	14	− 5	−70	−72
Hopefuls	14	− 1	−57	−73
Secular optimists	9	−22	−32	− 1
Pessimists	−22	12	78	110
Diffuse	−12	10	82	37
N =	(885)	(360)	(31)	(93)

There may be conflicts in Irish-Jewish or Italian-British interactions when matters of ultimate world view are either explicitly or implicitly on the agenda of either one side or the other, just as there might be conflict between Italians and Jews if occupational motivation should be the issue. There are no clear-cut distinctions on these matters that follow strictly religious lines. The differences are purely ethnic. When one comments on basic world view, occupational value, as well as on political style and family structure, one cannot be content with religious generalizations. It is not merely that Catholics generally seem to fall somewhere between Jews and Protestants, but that some Catholic groups seem to be rather more like Jews and other Catholic groups rather more like Protestants—and which group is more like another group often depends on the subject on the agenda.

The matter of values in American society is complex. Most Americans share similar moral, familial, occupational, and ultimate values. But the sharing is not complete. Within the common culture there is considerable variation in how one handles practical moral problems, occupational goals, and critical life situations. However, this variation on matters of value does not admit of easy analysis in religious terms, much less of convenient stereotyping. In handling specific moral issues,

TABLE 12.6

Ultimate Values of Catholic Ethnics
(Studies 5046 and 4172)

	PERCENT				
	BRITISH PROTESTANT	IRISH	GERMAN	ITALIAN	POLISH
Religious optimists	18	13	16	16	18
Hopefuls	26	23	17	12	17
Secular optimists	19	17	16	14	26
Pessimists	20	30	36	40	28
Diffuse	16	17	15	18	11
N =	(175)	(172)	(167)	(206)	(119)

Catholics seem to occupy a middle group between Protestant rigidity (or rectitude, if you wish) and Jewish permissiveness (or flexibility, if you wish). In occupational values, Irish and German Catholics have more in common with Protestants than they do with Polish and Italian Catholics, while in ultimate world view the Irish are more like the British Protestants, and the Italians are more like the Jews. The conventional wisdom of religious stereotyping on value issues is relatively easy to demolish, but the fluid pluralism that replaces it cries out for further analysis.

In Chapter 10, we were able to push beyond a mere description of fluid pluralism and show a link between diversity of family structure and diversity of political styles; we cannot do so with diversity of values because the data available simply do not permit further analysis. But we have shown, I think, that future investigation of values, either practical moral or ultimate religious, in American society ought to occur within a religio-ethnic and not merely a religious paradigm. Every generalization about values that begins with the word "Catholic" is likely to be misleading, if not erroneous, precisely because the generalization will mask substantial differences in values that exist among the Catholic subpopulations. It would be much more constructive for social commentators and researchers to desist from generalizations about Catholic fatalism and other-worldliness and begin to investigate the fascinating value paradoxes which seem to persist in American society. The Irish, for example, are both fatalistic and trusting; the Italians can combine deep pessimism with strong economic ambition, while the Jews, who share the Italian pessimism, are much more concerned about the inherent meaningfulness of their work. The actual pluralistic world of American value diversity requires not various shades of gray but the many hues of technicolor.

REFERENCES

1. Erling Skorpen, "Catholic Loyalty versus Protestant Morality? A Reader's Response," *Christian Century* 92 (October, 1975), p. 853.
2. Robert McNamara, "Organization, Colleges, and Values," *Sociological Analysis* 30 (Fall, 1969): 125–131.
3. Lawrence Kohlberg, "Development of Moral Character and Moral Ideology," in M. Hoffman and L. Hoffman, eds., *Review of Child Development Research*, vol. 1 (New York: Russell Sage Foundation, 1964), pp. 383–421.
4. William C. McCready, with Andrew M. Greeley, *The Ultimate Values of the American Population* (Beverly Hills, Sage Publications, 1976).

CHAPTER

13

The Catholic
Social Ethic

CATHOLICS differ from other Americans, and Catholic ethnics differ among themselves, as we have learned in the last several chapters. They are different in their ultimate world views, moral values, personalities, and family structures. But there have also been hints in my discussion that there may be an underlying "Catholic ethic" that takes a rather different view of the relationship between individual and society than other ethics currently influential in American life. Gerhard Lenski suggested[1] that the economic and educational achievement of Catholics could be impeded by their strong neighborhood family ties. We have seen that the ethnic group with the strongest family ties, the Italians, has prospered in a relatively brief period in the United States, and we have also seen that strong neighborhood involvement, far from impeding the economic achievement of parochial school graduates, seems to facilitate it. Lenski assumed that close neighborhood and family ties were an obstacle to social and economic progress. Modern capitalism, after all, is designed for a rational individual who vigorously and aggressively pursues his own enlightened self-interest. This model of man in society is fundamentally a mixture of Hobbes and Bentham. The fierce but enlightened self-interest of the highly competitive, atomized individual makes for both personal and social progress. Clearly such a model was implicit in Lenski's judgment about the deleterious economic effects of the Catholic ethic. Lenski conceptualizes the proper relationship between individual and society differently than do the Catholics about whom he writes. That he puts a higher value on his own model, if not for itself, than for economic achievement, seems never to occur to Lenski.

Yet strong familial ties have not interfered with Catholic educational and economic achievement, and strong neighborhood support seems to reinforce such achievement. Furthermore, the Catholic political style underemphasizes the "civic" approach and overemphasizes the direct personal contact approach (except for the Irish, that highly politicized people, who emphasize all possible forms of political involvement). But this political style has not prevented Catholics from achieving power in large American cities, nor has it necessarily meant that their cities are run less effectively than other cities. The ethnics' emphasis on family ties, particularized political contacts, and neighborhoods does seem to represent a social style relatively different from that which one finds among American Jews and Protestants (especially the Protestants). The Catholic style does not seem to be functionally inferior. It works. It may not be as esthetically or morally elegant as the other styles, but that should not be the issue. However, in much of the discussion of Lenski and other writers, it has been taken for granted that because the Catholic social ethic emphasizes the particular, the familial, the local, it is not only intellectually inferior, it is also inevitably less efficient. The data thus far suggest otherwise about its supposed inefficiency, and its intellectual inferiority depends on fundamental assumptions about human nature and the direction of evolutionary progress which are difficult to contend with. But at least the assumptions can be clarified and distinguished one from the other, and in this process we can learn the differences between Catholics and other Americans in matters of social style.

Is there such a thing as a Catholic social ethic? We have already noted the emphasis on family, neighborhood, and direct personalized contact. Three other researchers, Edward Laumann, Terry Clark, and James Q. Wilson, have assembled evidence which, when pieced together with the evidence already reported in this book, at last tentatively justifies some attempt to explicate a Catholic social ethic.

In careful sociometric studies of both Germany and the United States, Professor Edward Laumann of the University of Chicago has discovered that, with all other pertinent variables taken into account, Catholic friendship patterns curved like a closed circle, while Protestant friendship patterns radiated outward like spokes on a wheel. Commenting on this phenomenon, Laumann wonders whether it may represent a fundamentally different approach in world view.[2] If Laumann's careful work can be sustained in other communities besides the one he studied, one could argue plausibly that behavior patterns learned early prompt a proclivity among Catholics to form communal groups.

James Q. Wilson has the following extraordinary excerpt from an interview about the behavior of Irish Catholic officers:

Of course; no question about it. I have often thought, half-seriously, if you want to change a police department rapidly and effectively by putting new men at the top who will be loyal to the commissioner and do everything by the book and according to standard operating procedures, you could just about throw away the elaborate personnel tests and screening procedures we have devised and simply promote the northern European officers—the Germans, Scandinavians, English, and the like. You would get the same desired result with less time and money. [Q: Why is that?] It's hard to explain. It's not that they are necessarily any more honest than the Irish Catholic officers or that they are any smarter. It's more that they have a much greater and more obvious commitment to some set of rules, standards, or general principles as a way of doing and seeing things. The most striking fact about the Irish Catholic command officers is this department is the extent to which they rely on personal loyalties and the exchange of personal favors as a way of doing things. If there is a perfectly legal, routine way of doing something, you can almost be certain that many of your Irish Catholic officers will prefer to do it through some informal means instead. They deliberately step outside the formal system to do things informally. There is often, in fact usually, nothing at all wrong with what they are doing. It is just that they seem to feel more comfortable working through "contacts," intermediaries, and friends.[3]

Wilson suggests that the reason for this behavior might be the attitudes acquired by the Irish during the penal times when the indirect, the informal, the casual were the only ways to escape British tyranny. It is also true, however, that the Irish language is an extraordinarily gentle one, avoiding abrupt, dissonant words and phrases.* Finally, it may also be that whatever the past explanations, the Irish police officers have a much better feel for how to get things done by avoiding a rigidly structured formal bureaucracy. If you don't put anything on the record, and if you don't let anything go through official channels, you have a lot more flexibility. Maybe some of those sergeants should be made vice presidents for administration of corporations that suffer from bound muscles, hardened arteries, and diarrhea of the organizational charts.†

* There are no single words for "hello" and "goodbye," for example. Substitutes are "peace to this house" and "Jesus and Mary be with this house."

† A friend of mine told me of a certain younger Irish bishop (in the old country) who, overwhelmed with work and tied down by the burdens of office, was losing touch with his priests and people. I urged my friend to gather together some of his colleagues and warn the bishop in the bluntest possible language that he was getting himself into trouble. "Ah, well," I was told, "sure, we'll let him know, but there will be no formal confrontation about it, let me tell you. We'll take himself out to dinner some night and have a grand time, and then, over the last cigar, we'll very gently tell him what the problem is. It'll take only a few wee words and he'll know what we're talking about." I think such an approach is not the worst available means of upward communication. Would that all bishops were so sensitive to "a few wee words."

Terry N. Clark has written a long article in which he attempts to explain the strong relationship between the proportion of Catholics in a city, especially Irish, and public expenditures.[4] The Irish, he says, are particularists, and particularism is a tendency "to treat persons in terms of personal characteristics and continuing social contacts." Clark contrasts particularisms with "universalism," which "is a tendency to adapt general rules (of law or medicine or whatever) to interpersonal situations; abstract principles are applied to specific 'cases' or 'clients' by the universalistic judge or bureaucrat."[5] Clark gives as an example the person who is unhappy with street cleaning who could call the department of public works, which would be a universalistic solution; or he could call his friend the precinct captain, which would be a particularistic solution. "Both could yield identical results." Professor Clark does not suggest that the Irish are the only Catholic group to have such a particularistic style of operation, nor would I suggest that the Irish have a monopoly on implicit and unself-conscious Catholic social theory. They merely happen to illustrate it.

The work of Laumann, Wilson, and Clark, of course, does not "prove" the existence of an unself-conscious Catholic view of human nature and society that is widespread in the Catholic population; but the work of these scholars does at least raise the possibility that the "high tradition" theorists (like philosophers Jacques Maritain and Emmanuel Mounier, for example, or the writers of papal encyclicals) are not deriving their social perspectives from abstract principles so much as they are commenting on "instincts" about human social behavior that Catholics absorb very early in the socialization process. Nor need these "instincts" be formally theological in the sense that they are derived from clear and explicit doctrinal propositions. I would suggest as a working hypothesis that the "low level" tradition of Catholic social theory takes its origin from primordial intuitions about the nature of human nature and the nature of human society which have been part of the Catholic tradition for centuries. The police sergeant, the politician, the union leader, the bishop in a multiethnic diocese, the pastor in a multiethnic parish are all equipped with a set of templates for organizing, interpreting, and responding to behavior which have been carried along in the Catholic tradition without formal awareness for a long time.

Authors like Laumann, Wilson, and Clark are sophisticated enough to suspend judgment on whether the closed circles of friendship, the informal style, the "nonideological particularism" are good or bad. Other commentators are much less restrained. Orlando Patterson, for example, has raged against "particularism" and suggests that universalism is the only way for rational humans to live. One must break free of all the constraints of family, locality, and ethnic group to become a member of the universal society of educated human beings.[6] Professor Patterson is stating explicitly, I think, certain notions of evolutionary progress which are taken for granted by many American academics and which underpin much undergraduate and graduate education. To educate a young person from this perspective means to alienate him, to deracinate him, to "free" him from all the irrational, regressive, romantic ties of family, neighborhood, ethnic group, religion, and region. Whether such a goal is desirable or even possible is beyond discussion. Professor Patterson feels no need to prove the superiority of universalism; it is a matter to be taken for granted. Nor does he have the slightest hesitation about the long-range drift of humankind in such a direction. Again, one need not prove this; it is a self-evident fact.

The contemporary social science mentality, and hence the mentalities of many well-educated Americans, has been shaped by the image of "modernity," a mixture of Darwin, Marx, Weber, and Durkheim. The "modern," that is to say, the rationalized, formalized, universalized, bureaucratized, is inevitably triumphing over the archaic—the local, the particularistic, the informal, the nonrational. Such a triumph is not only a description of what is in fact occurring, it is also a prescription for what ought to occur. The tribal, the local, the sacred, the particularistic ought to vanish because they are "irrational." There is circularity in the argument, but arguments about basics always are circular. Even if one were to prove, as one can, that there are still vast residues of Gemeinschaft in the Gesellschaft society, one is simply told that these residues are merely the evidence of a temporary regression in the evolutionary process. The informal work group may still dominate production in factories, for example, but in the ideal computerized factory the informal work group will cease to have control. One can also argue that most human beings are a combination of the universal and the particular, and that even the most universalistic of academics get highly particularized about their students, their subdisciplines, their own factions, and fratricidal academic wars. Such arguments will be dismissed as *ad hominum*; the assertion that modern is good and more modern is better seems so self-evident to many Americans that discussion about the proposition is usually little more than the repetition of assertions. Faith in modernity has been failing in recent years, as is clear from the works of Heilbroner, Schaar, and Roszak,[7] but it is still strong. I suspect that a good deal of the elite anti-Catholicism in the country is based on the notion that the Catholic style and the Catholic ethic—perhaps even more so than Catholic theology—are inferior simply because they are not "modern." Thus, in an interview in the *Chicago Tribune*, Professor Morton Kaplan of the political science department of the University of Chicago could recently equate Catholic teaching with that of the Korean revivalist, Sun Myung Moon, under one common heading of "superstition."

Closed friendship circles, nonideological particularism, tribal (Irish) administrative style, close ties to family and neighborhood, personalized political behavior—they are all unmodern, inferior, regressive, and, in the final analysis, irrational and superstitious.

I shall attempt here a systematic description of what I take to be the Catholic social ethic, which is manifested in the various forms of political and social behavior already described in this chapter. I link this ethic with certain fundamental Catholic notions about the nature of human nature, which at its root is religious and theological. Such a reconstruction is obviously tentative and speculative. If this is how Catholics think about human nature and society, then the various forms of social behavior described thus far make sense. But there is no research evidence available, beyond that already cited, to confirm that many if not most Catholics are committed to these views of human nature and society to a greater extent than their non-Catholic counterparts. All one can say is that the pieces of evidence we have assembled thus far suggest that they might well be.*

1.   Paradoxically, Catholic Christianity has always been both more hopeful and more pessimistic about human nature than its Protestant opponents. Also, the Catholic is profoundly skeptical about remaking human nature through the manipulation of social structures. Salvation is possible, of course; indeed, it is readily available; but the *metanoia* of the Christian, however much it may be facilitated by some social structures and impeded by others, is ultimately a matter of internal personal conversion, and not of the modification of the basic traits of the human personality. In the Catholic view of things, one does not cease to be wary or afraid of other human beings; rather, one loves them despite one's fears. But while Catholicism is skeptical and reserved on the subject of remaking human nature, it takes a much brighter view of humankind in its present condition than does Protestantism. Humankind is deprived, perhaps, but not depraved. There is more that is

---

* Two caveats are important: First, while I call this ethic a "Catholic" ethic, it does not follow that Catholics have a monopoly on it. I might also have called it an "archaic" ethic or a "communal" ethic, finding parallel theories in Buddhism, the Federalist Papers, in some Jewish social thinkers, and in some non-Marxist socialists like Proudhon and his contemporary disciples. The ethic I discuss is called "Catholic" simply to indicate that Catholics seem to operate with this social view, and not to indicate that they have any monopoly on it.

Second, Catholicism as an institution has not infrequently ignored its own ethic and become in some ways far more authoritarian and centralized than its principle of subsidiarity would permit. It was not altogether facetiously argued in my seminary social ethics courses that the Catholic church was the first one to talk about pluralism and the last one to practice it, the first one to preach a living wage and the last one to pay it, the first one to insist on decentralization and the last one to allow it to occur within its own structure. Still, of all multinational organizations, the Catholic church probably does permit wider local decision-making powers (in the individual parishes, for example) than does any other.

admirable in man, as Camus said, than is contemptible. The flesh wars against the spirit, and the outcome is generally a toss-up, but Catholicism is inclined to bet on the spirit. Humankind is not fundamentally destructive, selfish, or individualistic. There is a selfish, aggressive, destructive aspect to our nature, but there is also a generous, trusting, and cooperative aspect to it, and Catholicism is more aware of and has far more confidence in that aspect of the human personality than does Protestantism. While under no illusion about the present condition of human nature, Catholicism still sees a sufficient amount of goodness and generosity to call people to virtue rather than to compel them to virtue.*

The Catholic social theory, then, makes the fundamental assumption that you create social order by an appeal to humankind's cooperative disposition, and not by force or by Hobbesian constraints. You sometimes have to back up pacific appeals by force, but society is not created by violence or maintained by force or even by formal social contract. It exists by definition when you have more than one social animal in the same physical environment.

2. Since it has a relatively more benign view of human nature, Catholicism can also take a more benign view of society. Social constraints that, in the Catholic view, exist to reinforce and support the cooperative tendencies of social beings are not so oppressive that one must keep them to an absolute minimum (as in the capitalist view of things), or impose them to a maximum amount in order that they may eventually wither away (the socialist view). Catholicism is under no illusion about oppressive states and oppressive societies; it has fought some and, lamentably, has allied itself with others; but it would deny that society or the state are necessarily or fundamentally oppressive. Society exists not so much to restrain human selfishness as to facilitate human growth (though in the process it must restrain some selfishness). And the state is not the arm of oppression to remake human nature or to keep the competitive system going; it is simply society's arm for ensuring an atmosphere of peace and tranquillity in which flawed but basically good human beings can create and share common enterprises and activities by which they may stumble through life a little more easily. Such a view of state and society is relatively modest. The state is neither the necessary evil of capitalism nor the temporary evil of socialism, but a mixture of good and evil, like everything human; under the proper circumstances, it is often more good than evil.

There is about this Catholic view of society and of the state something which may seem just a bit cynical. It is certainly utterly lacking in

---

* At least in theory. In practice, of course, many Catholic leaders over the last two millennia have attempted to compel virtue. In so doing, however, they have been completely false to their own theory. As Thomas Aquinas tells us, virtue is acquired by frequent repetition of *free* acts.

idealism, but it is also relatively free from disillusionment. The Catholic social theory expects more of both humankind and human society than does the capitalist, and substantially less than the socialist. It never expects to see a paradise on earth, but neither does it feel the need to remain content with life in the jungle. It has been around for a long time; it modestly expects to be around for a long time to come.

3. The Catholic social ethic is in utter horror of "scientific" or "rational" economic progress. It does not believe in Adam Smith's rational economic man. You never encounter people like that in the villages, the parishes, the neighborhoods, or, the Catholic social theorist suspects, anywhere in the world. Neither does the theorist believe in economic man's first cousin, the "homo faber" of the large corporate bureaucracy. You conceptualize a human being, the Catholic social theory argues, as a set of economic needs or productive skills only if you fundamentally misunderstand what human nature is all about. Furthermore, if you organize human enterprise in such a way that your theory and your structure assume that you have combined a group of productive skills for the maximum efficiency of output, you are not only likely to do terrible things to the human persons involved, you are also likely to defeat your own goals in the long run. You may try to ignore, eliminate, or even destroy the "nonrational" and "nonscientific" residues which cling to your corporate bureaucrat and industrial worker. You can take him out of his village and put him on a collective farm, you can move him from Rotterdam to Rochester and back, you may even get him to parrot an abstract universalistic ideology of the left or the right; but in the real world you and he will still be caught in the web of intimate, informal, particularistic, diffuse relationships rooted in biology, propinquity, or shared beliefs. The sensitive capitalist and socialist will say that of course they understand that the human being is a total person and not just a collection of economic needs or productive skills, but "rational" or "scientific" organization only requires economic needs and/or productive skills. One does not deny the other aspects of personality, and one may be peripherally interested in doing what one can to see that other needs are satisfied, but when push comes to shove, there is no particular reason to think that these needs are important either in satisfying the demands of the profit-hungry stockholders or the goals the planning board set up for the five-year plan.

The Catholic social theory obviously dissents. Human life is both an organic and differentiated unity. Human beings live in dense networks of overlapping commitments, relationships, loyalties, involvements. There is a pluralism of relationships in human society that constitutes, in Maritain's words, "organic heterogeneity in the very structure of civil society."[8] To ignore that organic heterogeneity even for the purpose of drawing up an organizational chart is not only to do violence to the people whose jobs appear there, but also to risk the ultimate collapse of one's effort, because abstractions are not to be found on the assembly line or in offices of the bureaucracy. One cannot abstract human beings caught in relationships, committed in loyalties, and absorbed in beliefs, viewpoints, and prejudices. To forget that is to misunderstand everything.

4. The Catholic social theory, then, has a profound respect, one might say reverence, for the informal, the particular, the local, the familial. It does not believe that this delicate and intricate web of primordial ties which binds human beings together in dense and close relationships should be ignored or eliminated. It does not believe that evolutionary progress is moving the human race from one end of the Parsonian pattern variables to the other end. It does not believe this for two reasons. First, its definition of human nature as social and relational does not permit it to see intimacies as temporary or transient or easily replaceable. Second, in all its dealing with the ordinary people who constitute its congregation, Catholicism encounters very few who are suitable for the life in the temporary society. With rare exceptions, those transient citizens who can decamp from Rochester to Rotterdam to Riaydh, or from Cambridge to Hyde Park to Berkeley at a moment's notice are unhappy and frustrated. Catholicism doubts that they are the leading edge of evolutionary progress—and it is skeptical that many human beings would want to live that way unless they were forced to.

The American church in particular (if it ever bothers to reflect on itself) would be astonished at the way the peasant villages of Galway and Mayo and Sicily and Bari were reconstructed in the slums of New York and Chicago and Boston—perhaps they were not so much reconstructed as simply transplanted.

The Catholic social ethic does not believe that you should try to destroy the web of personal intimacy that every human being spins around himself. More than that, it believes that you cannot destroy it. Abolish Gemeinschaft, if you will, but it will reassert itself in the giant factory or the mammoth collective farm. Before you know it, it will be in control again, and its influence will be all the harder to deal with because your ideology forces you to deny its existence.

This argument is perhaps the most telling point that the Catholic social theory can make against both socialism and capitalism, for it is both an ethical and an empirical point. You should not try to destroy Gemeinschaft, says the Catholic social theory, and, besides, you cannot. It is a judgment which admits of empirical falsification or verification, and, as we shall see shortly, the evidence is overwhelmingly against those who think you can dispense with or ignore the primordial, the particularistic, the local, the intimately interpersonal.

5. The Catholic social theory categorically rejects the notion that one can or should sacrifice the present for the future. It is outraged that anyone should suggest that you should. The end does not justify the means, however noble the end. A future material paradise of peace, prosperity, plenty, and freedom does not justify slave labor communes, hunger, misery, torture, and repression. The Catholic social theory takes this position in part because of its view of the nature and destiny of the human person created in God's image and called by Him to grow in knowledge and love. A creature who hungers for the absolute in the roots of his personality is not to be a pawn in a planner's program for economic development, for the Catholic social theory also knows that this benign future paradise cannot be counted on to appear. The new class, like the old class, converts means into ends, and becomes much more interested in preserving its own power, privilege, and prerogative than in producing the new order to which it is allegedly committed. With the observation of its own clergy and hierarchy succumbing to the temptations of power, Catholic social theory is under no illusion that other powerful bureaucrats, no matter how pure their original motivation, can ultimately escape the same temptations unless there are strong and effective legal and structural limitations on their power. The Catholic social theory knows enough about grand inquisitors to be suspicious of them. You do not build a better future by exploiting the present. The "pyramid of sacrifice" strategy really only provides a rationale for staying in power; it is both immoral and self-deceptive.

6. Because of its beliefs about human nature, about the organic and heterogeneous pluralism of human society, about the critical importance of the web of primordial relationships in which each person is enmeshed, and because of its convictions about fundamental dignity and the value of each individual person, the Catholic social theory must be profoundly skeptical of all attempts at "modernization" or "development" which purport to improve the material lot of a people by destroying its culture and its social structure. Doubtless some cultures and structures are more open to growth and development than others. Doubtless, too, there must be changes in some societies if poverty and hunger are to be eliminated; but the Catholic social ethic has too much respect for the tenacity of tradition, the power of custom, the resistance of inertia, and the strength of fundamental belief and values to think that a collective farm, a multinational corporation, a giant steel mill, or a fleet of tractors can enjoy anything but short-range success when imposed on a people who view these innovations as fundamentally alien to their way of life. If you can integrate innovation into the culture and structure of a community, then the innovation may prosper, but if you tell the people to jettison their values and forget their customs and embrace the new technology for some such goal as increased food

production, they may elect for customs and values in preference to food (much to the astonishment of the Peace Corps volunteer, the socialist bureaucrat, or the industrial technician). You may dispense medical services in the Himalayas, New Delhi, and the southwest side of Chicago, and in each place you may use the most modern and advanced techniques of preventive and curative medicine; but you must never forget that you are delivering your services to an individual who is not an isolate, bereft of values and convictions. You are delivering them to a person who has value and worth in his own right and who is caught up in an intricate web of human relationships, commitments, and loyalties. Any attempts at social change, the Catholic theorist must argue, which disregard these facts about human nature are both morally and intellectually wrong—and won't work besides.

7. Finally—and this may be the easiest way to tell the Catholic social theorist from the socialist and the capitalist—he is profoundly suspicious about size. Just as the capitalist in principle and the socialist in practice thinks larger is better, so the Catholic theorist can only respond that small is beautiful. Of course, one must avoid the romantic temptation to assume that small is invariably beautiful or better. A hundred people cannot support an airline, a country of small farmers cannot produce enough food to feed the world—or even one large industrialized society. "Too big" is not an absolute matter but a relative one, and from the perspective of the Catholic social theory, something becomes too big when it is bigger than it has to be to do the job. The principle of subsidiary function is perhaps the central theme of Catholic social theory. It vigorously argues that nothing should be done by a larger organization that could be done as well by a smaller one; and nothing should be done by a higher bureaucratic level that can be done just as well by a lower level.

Perhaps the mistake of much Catholic social theory in the past was to argue as though the principle of subsidiarity was a philosophical deduction rather than an empirical observation. You keep things as small as you possibly can because they work better that way, there is more flexibility, better communication, more room for innovation, adaptation, and quick response to new problems. You decentralize decision making as much as you can so that those who are responsible for carrying out decisions can participate in the decision-making process, making their motivations to achieve successful implementation much stronger. These are not merely ethical principles; they are empirically documented facts. Such facts are ignored today in the organization and administration of corporate bureaucracy, which merely proves that the blinders of ideology and habit can filter out critically important information. In the short run, giantism is efficient. You can maximize production quickly with "economies of scale." The only trouble is that the corporate organization is made up of more than just machines; it is also composed of human beings, and in the long run, economies of scale easily lead to uneconomies of human effectiveness. This fact has been proven time and time again, but it still does not seem to have entered the thinking of the theorists of either corporate capitalism or corporate socialism.

Catholic social theory would also argue that no matter how large the organization, it is immoral, erroneous, and foolish to treat it as though it were made up of atomized, isolated individuals. Even if you bring a group of complete strangers together to operate your plant, those strangers will set up informal social networks during the course of the first morning. Soon *they* will run the factory, not you. You may drive the peasant out of his old village and set him up in a clean, new, efficient agricultural commune; you may even threaten him with death if he doesn't live up to your standards of productivity; but that does not mean friendship networks will not emerge to sabotage the goals set for you by the Central Planning Board in Peking, Havana, or Moscow. From the perspective of Catholic social theory, it is not a question of an organization without Gemeinschaft but rather an organization which recognizes and works with it.

The Catholic social theory differs from capitalism and socialism, for example, in its view of individual and class conflict. Capitalism and socialism assume that in the natural state of things individuals and classes are in conflict. In the capitalist society the state is the organ of the ruling class (though capitalists would be reluctant to admit that quite so explicitly); in a socialist society the state allegedly becomes the instrument of the oppressed class against the ruling class (though in fact it usually becomes simply a tool of the new class). Both theories are uneasy about conflict within their own societies. Social and political unrest, or even "too much" diversity, is viewed by the capitalist as a threat to the stability of his society and the maintenance of high levels of productivity. The socialist, once he has gained power, considers political opposition to be "counterrevolutionary" and vigorously represses dissent. Both socialist and capitalist applaud conflict and competition in theory and do their best to repress it in practice—perhaps because both are impressed by the inherently unstable nature of human social institutions.

The Catholic theory, on the other hand, is much more relaxed about the stability of human institutions because it views them as based on the dense and intimate interpersonal networks of "lesser groups" which it takes to be the raw material of society. Since these "lesser groups" (family, local community, friendship circle, local church, neighborhood) are normally more cooperative than competitive, the Catholic social theory assumes a matrix of much greater social cooperation than do its two individualist social theory adversaries. Of course, even within much lesser groups there is competition (between husband and wife, parents and children, power players and bridge players), though the competition rarely destroys the cooperative structures. Catholic social theory assumes that such competition is normally more healthy than not, and is not greatly disturbed by it.

Precisely because it has much greater confidence in the positive and constructive forces that are at work—or at least can be at work—in society, the Catholic social theory is less worried about the society's capacity to deal with diversity, competition, and conflict in its "greater groups." It realizes that such competition can lead to instability and societal breakdown, but that it does not necessarily do so. Hence Catholic social theory, unlike its socialist counterpart, sees no reason to repress dissent (however much it may have been repressed in certain Catholic states). Unlike capitalism, Catholic social theory is unworried that conflict between labor and management, for example, will destroy the industrial enterprise. After all, the Catholic theorist notes, labor and management do have common interests as well as diverging ones.

But if it takes conflict and competition for granted (even viewing them as good), the Catholic social theory is deeply suspicious of those who deliberately set out to stir up conflict between classes—especially when this conflict seeks to make the opposing class the scapegoat, the enemy to be destroyed, the evil cause of all one's trouble. Thus the Catholic theorist must reject in principle the current epidemic of romantic scapegoating of certain "oppressor" groups—men, whites, older people, the Northern Hemisphere. For all his respect for societal networks, the Catholic theorist is absolutely committed to the worth and dignity of the individual person, and cannot tolerate the arbitrary assignment of guilt or nonvalue to anyone because of a characteristic acquired at birth. Furthermore, such scapegoating ignores the mutuality of interests that such falsely opposed groups obviously have. Finally, the Catholic theory observes that in the "lesser groups" the normal state of relationships is a mixture of cooperation and competition, with the former predominating for the most part. To pretend that such relationships are not the raw material of all society seems to the Catholic social theorist to be foolish posturing.

Out in the parishes there is competition between men and women, between young and old, among various ethnic groups. It has ever been thus, and the Catholic theory accepts that it always will be. Yet, for the most part, the individuals involved relate to one another with tolerance, affection, and even love. You can compete with people, the Catholic social theory observes, and still love them. In fact, love and conflict are correlates as well as contradictions. Those mass movement leaders who think you can have effective conflict and competition only when you stir up hatred against an opponent (or class) which must be destroyed are quite literally, in the view of the Catholic theorist, doing the work of the devil. Furthermore, in the long run their strategy won't work, for it is not natural for women to view men as enemies *all the time*, or for the young to view their parents as enemies *all the time*, or even for one group to view another as enemies *all the time*.[9]

The Catholic social theory is not derived solely from abstract principles. Like any good theory, it is the product of both induction and deduction, of reflection and experience, the reinterpretation of reflection in general propositions and the application of those propositions for testing and verification in the practical order. In the past, much emphasis was placed on the deduction of a Catholic plan for restructuring society from St. Thomas, the papal encyclicals, or a frequently romanticized version of what life was like in the Middle Ages. I have tried here to rely more on analysis of and reflection about the Catholic grassroots, unself-conscious experience of a social theory. While I wrote this chapter I wondered whether I was reflecting on the "high tradition" of Catholic social theory as contained in the encyclicals and writers like Maritain and Mounier, or whether I am reflecting on my own personal experience and articulating a world view that I had before coming into contact with the worthy philosophers. I incline to the latter explanation. Like the Catholic labor leaders of the 1930s and 1940s (John Brophy, Philip Murray), I resonate positively to the social encyclicals because they articulate what I already "feel in my bones" to be true.

I think the "high tradition" of Catholic social theory is merely a formalization and articulation of the insights and presuppositions of the experience of the politicians, union leaders, parish priests, and immigration era bishops who responded to their practical problems in an American urban environment out of instinct, with precious little need to refer to Leo XIII (whom most of them never read) or Maritain (whom most of them never heard of). Implicit in such an assertion is the notion that Catholics acquire very early in life, in the intimacy of the childhood socialization process, a fundamentally different view of human nature and society.

If the reader wonders whether the Catholic social theory is reactionary or radical, left or right, conservative or liberal, I would observe that such questions can be asked only if one makes the rigid assumption that the available alternatives for organizing human life in the modern world are capitalism and socialism. If one cannot see beyond them, the Catholic social theory is obviously irrelevant. It is surely reactionary in the sense that it rejects the idea that evolutionary progress has eliminated the primordial and the particular from the human condition. It is also reactionary in that it suggests that heterogeneity, particularism, and primary group relationships are not to be ignored by social planners, policy makers, and administrators. But it is profoundly radical in that it rejects the basic and fundamental assumptions on which both capitalism and socialism are based, and the perspective within which the capitalism versus socialism debate occurs. It says to Marx and Adam Smith, "You are both wrong."

It is also conservative in the sense that it believes in conserving such fundamental institutions as neighborhood, family, village, community; but it is liberal, indeed radically liberal, in its demand that human beings be freed from the oppressive chain of the large corporate bureaucracy (and it doesn't care which ideology underpins the bureaucracy.)

Little if any empirical evidence has been collected to establish that there is indeed a Catholic social ethic of the sort described in this chapter. One of the reasons for the absence of evidence is that social science funding agencies are not yet persuaded that the matter is of sufficient importance to investigate. After all, what difference does it make if one-quarter of the population has rather different social values than the other three-quarters? However, in an analysis of data collected by my colleagues Kathleen McCourt and David Greenstone,[10] I was able to test directly for the presence of some sort of "neighborhood ethic" among Catholics.

If indeed Catholics are disproportionately likely to have convictions about the importance of dense and organic networks of informal personal relationships, one would expect them to be more deeply attached to their cities and their neighborhoods, to know more neighbors in their neighborhoods, to share neighborhoods with members of their families, to have high proportions of their close friends in the neighborhoods, and to spontaneously describe "friendly people" as what they most like about their neighborhoods. All of these suspicions (I'll not glamorize them by calling them "hypotheses") were sustained by the evidence (Table 13.1). Catholics do have a stronger loyalty to their neighborhoods, a loyalty which we would have predicted on a priori grounds from our knowledge of the style of religious reflection of the Catholic high theorists and the Catholic social and political ideology as it has been described in this chapter. Of course, anyone who has ever lived in a neighborhood would have taken such findings almost for granted.

It is a long way from an Italian calling a precinct captain about garbage collection, a Pole refusing an academic appointment in another city because he does not want to leave his neighborhood, and a Slovak reacting as though a threat to his neighborhood is a threat to his very self, to such high-level theorizing about the nature of human nature and

TABLE 13.1

Protestants, Catholics, and Neighborhoods *

	PROTESTANTS (N = 308)	CATHOLICS (N = 587)
Born in same city	53%	79%
Not planning to move	54%	68%
Move outside SMSA	31%	23%
	(of those who are moving)	(of those who are moving)
Like neighborhood	66%	74%
Know "many" neighbors	53%	62%
Some family in neighborhood	25%	38%
More than half of close friends in neighborhood	16%	25%
"Friendly people" most liked neighborhood attribute (spontaneously mentioned)	10%	15%

*The sample includes women active in community organizations, a control group of inactive women, and husbands of 50 percent of the sample women.

society. If it exists at all, the Catholic social ethic can be found in the concrete—not as an abstract theory but as a series of templates, a set of pictures and images for responding to concrete social situations and problems. There is some evidence that a young Catholic grows into adulthood with a somewhat different set of templates than the young Protestant or the young Jew. Most of the evidence for this, however, is indirect: Catholics engage in certain kinds of behavior that is frequently "unaccountably" different. Such differences suggest, though they do not prove, the existence of a different set of pictures. Anyone who wishes to understand better the complexities of American society ought to feel constrained to learn more about the possible existence of this different set of templates. It is indicative of the present transitional state of American Catholicism (about which more in the final chapter) that Catholics themselves are not engaged in such a search. Much of the Catholic intelligentsia are busy denying both to others and to themselves that they have a different set of pictures, and most of the rest of the Catholic population is still too busy working for more achievement and success to examine the possibility that it may be traveling up a rather different road from Americans of other collectivities.

REFERENCES

1. Gerhard Lenski, *The Religous Factor*, rev. ed. (Garden City, N.Y.: Doubleday Anchor, 1963), p. 345.

2. Edward O. Laumann, *Bonds of Pluralism* (New York: Wiley-Interscience, 1973).

3. James Q. Wilson and Edward C. Banfield, "Political Ethos Revisited," *American Political Science Review* 65 (December 1971): 1048–1062.

4. Terry Nicholas Clark, "The Irish Ethic and the Spirit of Patronage," *Ethnicity* 4 (December 1975): 305–359.

5. Ibid., p. 17.

6. Orlando Patterson, *Sociology of Slavery* (Rutherford, N.J.: Fairleigh Dickinson Press, 1970).

7. See Robert Heilbroner, *An Inquiry into the Human Prospect* (New York: Norton, 1974); John Schaar, "Reflections on Authority," *New American Review* 8 (1970): 66–68; and Theodore Roszak, *The Making of a Counterculture* (Garden City, N.Y.: Doubleday Anchor, 1969).

8. Jacques Maritain, *True Humanism* (London: Geoffrey Bles, The Centenary Press, 1938), p. 157.

9. See Rosemary Harris, *Prejudice and Tolerance in Ulster: A Study of Neighbors and Strangers on a Border Community* (Totowa, N.J.: Rowman & Littlefield, 1972).

10. J. David Greenstone and Kathleen McCourt, "Politics and Families: Changing Roles of Urban Woman." Ongoing study supported by the Center for Studies of Metropolitan Problems, National Institute of Mental Health, under research grant 1 ROI MH 23786, at the National Opinion Research Center, Chicago, Illinois.

CHAPTER
14

The Emergence of
the Communal Catholic

THREE THEMES have run through this book. First, I have argued that there is cultural division of labor in American society by which Catholics are systematically underrepresented in certain high-prestige positions, most notably the foundations, the great private universities, the elite national media, and, to some considerable extent, the boards of the most powerful business corporations. This cultural division of labor, I have suggested, has been legitimated by a mythology which views Catholics as narrow, rigid, antiintellectual, morally authoritarian, politically corrupt, economically unsuccessful, politically conservative, racially prejudiced, and tied down by the obstacles of family and neighborhood loyalty. The stereotype of the "blue-collar ethnic," a racist, reactionary, hardhat, summarizes this mythology. I have examined the various components of the stereotype and found that there is no empirical evidence to support any of them, that on the contrary, the overwhelming weight of evidence goes in the opposite direction. The educational and economic achievements of Catholics have surpassed that of Protestants in the metropolitan regions of the North, and younger Catholics are pursuing careers of scholarship and research at the best state universities in the country. Politically they are moderate left, especially on the traditional New Deal issues; racially they are less opposed to integration than comparable Protestants, and Irish Catholics are right behind Jews in their support for racial integration. They were more likely to oppose the Vietnam War from the beginning than were Protestants, and have defected neither from the Democratic party nor from regular patterns of electoral support for Democratic candidates. They are more flexible in handling moral decisions than Protestants, and while they differ in family structure and political style from other Americans, the diversity within the Catholic community in these more personal values and behaviors is, if anything, greater than the differences between Catholics and Protestants or between Catholics and Jews.

However, the mythology still does underwrite a cultural division of labor. Our data showed that discrimination against Catholics occurs not at the bottom end of the economic ladder but at the top, where their educational and financial success does not seem to have gained them access to the most prestigious occupational positions.

The anti-Catholic mythology is not nearly as pernicious as racist, anti-Semitic, or sexist mythology. It does little direct harm to the overwhelming majority of American Catholics, affecting directly only those well-educated Catholics who aspire to enter into positions in the national, elite-dominated institutions. Indirectly, the mythology may have an impact on many more American Catholics, in that it both indicates and justifies policy decisions to make urban Catholic neighborhoods the targets for social programs designed to correct injustices against other minority groups. It is right, the mythology says, that Catholics should bear the costs of social change because of their own bigotry and general inferiority.

The second theme is the "mosaic with permeable boundaries" or "stewpot" picture of American pluralism. This model assumes that American ethnic groups are less a residue of Old World culture and more a dynamic innovation of the American environment. Ethnic groups, I would suggest, have a momentum and life history of their own, leaning sometimes toward assimilation and at other times toward more highly self-conscious differentiation. To put the matter more specifically, it is in the nature of the stewpot model that in some ways members of American ethnic groups may be becoming more like one another, and in other ways more different. Thus, at the present time, American Catholics have achieved economic and educational parity—indeed superiority—when compared with American Protestants, but they continue to maintain differential familial structures and values, and loyalty to their religious tradition.

The argument that Catholics could not achieve economic success because their values were fundamentally different from those of the rest of society, and that those few Catholics who did achieve success left behind their specific values, is clearly untenable. The evidence of this book shows that Catholics have achieved educational and economic success in overwhelming numbers, and have not rejected their family and neighborhood loyalties, political styles, or religious tradition.

Finally, I have drawn a sharp distinction between the Catholic collectivity or subpopulation and the Catholic church as institution. Such a distinction has not seemed very important hitherto, because both inside and outside the Catholic population the identification of the Catholic collectivity with the institutional church has been taken for granted. But as we have seen, in the wake of the Second Vatican Council and in particular as a result of the birth control encyclical *Humanae Vitae*, the credibility of the leadership of the institutional church has been badly damaged. Rejection of the church's sexual ethic and the decline of religious devotion on a number of indicators has not led Catholics to defect from the church in very large numbers. Indeed, the powerful loyalty of the American Catholic population to the parochial school system seems virtually intact. The weakening of the hold of the institutional church has had little to do with assimilationist pressures from the rest of society, and has resulted rather from the monumental incompetence of ecclesiastical leaders.

The most likely projection for the future is the emergence of a large group of what I have called elsewhere "the communal Catholic." They are loyal to the Catholic collectivity and at least sympathetic toward its heritage. At the same time, they refuse to take seriously the teaching authority of the leadership of the institutional church. Such communal Catholics are Catholic because they see nothing else in American society they want to be, out of loyalty to their past, and they are curious as to what the Catholic tradition might have that is special and unique in the contemporary world. There is no specific empirical evidence to document the emergence of the communal Catholics—though the decline of acceptance of the church's sexual ethic and the persistence of loyalty to Catholic schools is perhaps evidence enough of a new and self-consciously selective style of affiliating with the Catholic tradition. Furthermore, the various conscious manifestations of ethnic concern which have developed in the last decade, particularly among the third and fourth generations of eastern and southern European Catholics, is confirmation of the development of a communal style of affiliation with Catholicism. To be Italian American may be now not only a way of being American but also a way of being Catholic.

The communal Catholic is particularly likely to be found among those well-educated younger Catholics who find themselves on the fringes of the nation's intellectual and cultural elites, either trying to gain admission or conscious of the barriers that bar admission. It is by no means accidental that ethnic militancy has occurred not among the Catholic working class but among the Catholic intelligentsia, for they are the ones who are most likely to experience the negative effects of the cultural division of labor.

In an earlier day, Catholics who aspired to enter the nation's intellectual and cultural elites had to pay the price of abandoning their religious heritage, or at least hiding whatever tenuous affiliation they maintained. Such alienation was also frequently imposed by the Catholic community on those who sought to enter the intellectual and cultural elites. There was little room for intellectual or cultural concern in the immigrant garrison church of the 1920s, 1930s, and even the 1940s. The only way many young Catholics of those years were able to pursue their cultural and intellectual interests was to struggle out of the clutches of their Catholic past.

Now the battered, fluid, and pluralistic church of the years after the Vatican Council is able to constrain the freedom of no one. As one American sociologist put it to me, "I left the Church when I was twelve years old. At forty-five I woke up one morning to discover I was back in it—not because I had changed but because the church had moved its boundaries out so far that I was inside once again." While some Catholic scholars and artists are still busy fighting the old garrison church, the walls of the garrison have long since collapsed, and a substantial number of young Catholic intellectuals and professionals find nothing in the institutional church to justify fierce anger. Furthermore, as a result of the tremendous educational achievements of Catholics in the last thirty years, there are now simply too many young Catholic professionals for the nation's elite groups to exclude.

The communal Catholic, I would suggest, will have both a more selective and more self-conscious style of being Catholic. He will also be less likely than either the old neighborhood Catholic or the old Catholic intellectual to be unperturbed by the cultural division of labor and the anti-Catholic mythology that underwrites it. We will not see a militant working-class or even middle-class Catholic ethnic self-consciousness, for both the working-class and middle-class Catholics have relatively little to complain about in the United States, except in those areas where they have been chosen to pay the price for other people's guilty consciences. But as more and more Catholics move to the fringes of the intellectual, cultural, journalistic, and administrative elites and encounter barriers to their admission, they are likely to vigorously attack those barriers and the mythology which buttresses them.

The communal Catholic represents a new paradox of American pluralism. He will be less involved in the institutional church but will be more self-consciously, even militantly, Catholic. It is beyond the purpose of this book to speculate how the communal Catholic will work out in detail his relationship with both the organized church and the rest of American society. Specific events and the actions of leadership of the church and of the cultural elites will have unpredictable effects on the development of the communal Catholic ethos and style. But that the communal Catholic does exist and will continue to develop is certainly consonant with the evidence and with the model of American pluralism as a "mosaic with permeable boundaries."

But both the melting-pot model and the anti-Catholic mythology are still very much a part of the baggage that America's intellectual and cultural elites carry around with them for interpreting and understanding their own society. So I conclude this book on the same note of futility with which I began. Many of those for whom the book is designed will simply not believe it, either in sum or its individual parts. What's more, they will doubtlessly find Catholics who will reinforce them in their nervous confidence that they do not have to question their basic assumptions about American Catholics.

Far be it for me to challenge their stereotypes.

INDEX